Design, Synthesis and Evaluation of Highly Functionalized Peptoids as Antitumor Peptidomimetics

Zur Erlangung des akademischen Grades eines

DOKTORS DER NATURWISSENSCHAFTEN

(Dr. rer. nat.)

Fakultät für Chemie und Biowissenschaften

Karlsruher Institut für Technologie (KIT) - Universitätsbereich

genehmigte

DISSERTATION

von

Carmen Cardenal Pac

aus Barcelona

Dekan: Prof. Dr. Peter Roesky

Referent: Prof. Dr. Stefan Bräse

Korreferent: Prof. Dr. Hans-Achim Wagenknecht

Band 47
Beiträge zur organischen Synthese
Hrsg.: Stefan Bräse

Prof. Dr. Stefan Bräse
Institut für Organische Chemie
Karlsruher Institut für Technologie (KIT)
Fritz-Haber-Weg 6
D-76131 Karlsruhe

Bibliografische Information der Deutschen Bibliothek

Die Deutsche Nationalbibliothek verzeichnet diese Publikation in der Deutschen Nationalbibliografie; detaillierte bibliografische Daten sind im Internet über http://dnb.d-nb.de abrufbar.

ISBN 978-3-8325-3721-0
ISSN 1862-5681

Logos Verlag Berlin GmbH
Comeniushof, Gubener Str. 47,
10243 Berlin
Tel.: +49 030 42 85 10 90
Fax: +49 030 42 85 10 92
INTERNET: http://www.logos-verlag.de

Als pares i al Felip

Die vorliegende Arbeit wurde in der Zeit vom 10. Oktober 2010 bis zum 06. November 2013 am Institut für Organische Chemie des Karlsruher Instituts für Technologie (KIT) unter der Leitung von Herrn Prof. Dr. Stefan Bräse durchgeführt.

Hiermit versichere ich, die vorliegende Arbeit selbstständig verfasst und keine anderen als die angegebenen Quellen und Hilfsmittel verwendet sowie die Zitate kenntlich gemacht zu haben.

Table of Contents

1. Abstract

Peptides and proteins are involved in a diverse range of essential processes in living organisms, and, therefore, they are of great interest for medical and biological applications. However, their fast degradation by proteases and, consequently, their poor bioavailability make them unsuitable as therapeutic agents. Peptoids are structural isomers of peptides with improved proteolytic resistance, good cell penetrating properties and easily accessible and derivatizable synthesis. Therefore peptoids are promising peptidomimetics for therapeutic applications.

The aim of this project was the design, synthesis and evaluation of peptoid analogs of short amino acid sequences with antitumor activity.

A small library of peptoids was synthesized mimicking a pentapeptide that inhibits ligand dependent Met activation by competing with the coreceptor function of v6-containing CD44. In addition to functional group modifications, derivatives with reduced flexibility could be obtained by incorporating α-chiral submonomers, peptide-peptoid hybrids or cyclization. Many of the synthesized peptoids were able to inhibit HGF induced Met activation and block scattering and migration of tumor cells like the original peptide. Comparison of the inhibitory ability of the peptide and the different peptoid analogs led to the establishment of some structure-activity relationships. Moreover, the peptoid that exhibited the strongest inhibitory effect *in vitro*, was also able to reduce the number of liver metastatic events and decrease angiogenesis *in vivo*, holding high potential for cancer therapy.

Peptoid analogs of a sequence of the BAG-1 protein inhibiting tumor cell growth in prostate cancer were also prepared. While addition of the peptoids alone was not able to reproduce the effect of the protein, the fact that rhodamine-labeled analogs were able to penetrate the cell and partially colocalized with the receptor, opens the way to the development of new inhibitors combined with targeting sequences.

Furthermore, a method for the synthesis of amide-containing submonomers as free bases and their incorporation into peptoid oligomers has been described. Different-length homo-oligomers of α-chiral amide residues were successfully synthesized for structural investigations *via* circular dichroism spectroscopy.

Kurzzusammenfassung

Peptide und Proteine sind in lebenden Organismen an einer Vielzahl von essentiellen Prozessen beteiligt und sind daher von großem Interesse für medizinische und biologische Anwendungen. Allerdings sind sie aufgrund ihres schnellen Abbaus durch Proteasen und der daraus folgenden schlechten Bioverfügbarkeit ungeeignet um als therapeutische Wirkstoffe eingesetzt zu werden. Peptoide sind strukturelle Isomere der Peptide und zeichnen sich besonders durch ihre höhere Stabilität gegenüber enzymatischem Abbau, ihre guten Zell penetrierenden Eigenschaften, sowie ihre einfache und derivatisierbare Synthese aus. Daher sind Peptoide äußerst vielversprechende Peptidomimetika für therapeutische Anwendungen.

Das Ziel dieses Projekts war es Peptoid-Analoga von kurzen Aminosäure-Sequenzen mit Antitumor-Aktivität zu entwickeln, zu synthetisieren und zu evaluieren.

Im ersten Projekt wurde eine kleine Peptoid-Bibliothek synthetisiert, welche ein Pentapetid nachahmt. Dieses Pentapeptid fungiert als Inhibitor für die Liganden-abhängige Aktivierung von Met, indem es mit der Korezeptor-Funktion von v6-haltigem CD44 konkurriert. Neben Peptoiden mit modifizierten funktionellen Gruppen wurden auch Derivate mit reduzierter Flexibilität erhalten. Dies konnte durch die Einführung von α-chiralen Submonomeren, die Darstellung von Peptid-Peptoid-Hybriden oder durch Cyclisierung erreicht werden. Viele der dargestellten Peptoide besaßen die gleiche Aktivität wie das ursprüngliche Peptid und konnten die HGF-induzierte Met-Aktivierung verhindern, sowie das Streuen und die Migration von Tumor-Zellen blockieren. Durch einen Vergleich der inhibierenden Wirkung des Peptids und der unterschiedlichen Peptoid-Analoga konnten einige Struktur-Aktivitätsbeziehungen aufgestellt werden. Außerdem konnte das Peptoid, welches *in vitro* die stärkste inhibierende Wirkung besaß, *in vivo* sowohl die Anzahl der Lebermetastasen reduzieren als auch die Angiogenese verringern. Dieses Peptoid besitzt somit ein hohes Potential für die Krebs-Therapie.

Weiterhin wurden auch Peptoid-Analoga von einer Sequenz des BAG-1 Proteins, welches das Wachstum von Tumor-Zellen in Prostata-Krebs inhibiert, dargestellt. Die Peptoide waren jedoch alleine nicht dazu in der Lage den Effekt des Proteins zu reproduzieren. Allerdings konnten Rhodamin-markierte Peptoid-Analog die Zelle penetrieren und kolokalisierten zum Teil mit dem Rezeptor. Dies ermöglicht die Entwicklung von neuen Inhibitoren, welche mit zielspezifischen Sequenzen kombiniert werden.

Zudem wurde eine Methode entwickelt, welche die Synthese von Amid-haltigen Submonomeren als freie Basen und deren Einführung in Peptoid-Oligomere ermöglicht. Für Struktur-Untersuchungen mittels Zirkulardichroismus-Spektroskopie wurden unterschiedlich lange Homo-Oligomere mit α-chiralen Amid-Resten dargestellt.

2. Introduction

Peptides perform a wide variety of essential functions in living organisms with great specificity and precision. Their diverse roles include hormones, toxins, enzymes, antibodies, receptors, neurotransmitters, structural proteins, etc.[1]

However, the use of peptides for biological and therapeutical applications is limited, mainly due to their sensitivity to proteolytic degradation, which results in poor bioavailability. For this reason there is a strong interest in developing compounds that mimic the structure and/or function of peptides and proteins overcoming their drawbacks.

2.1 Peptidomimetics

Peptidomimetics were initially defined by Giannis and Kolter as "compounds that imitate or block the biological effect of a peptide at the receptor level".[2] One year later Gante extended the definition to include both peptide function and structure: "a peptidomimetic is defined as a substance having a secondary structure as well as other structural features analogous to that of the original peptide, which allows it to displace the original peptide from receptors or enzymes".[3]

An ideal peptidomimetic should be resistant to proteolytic degradation, improve the oral bioavailability of the original peptide, and increase its effectivity and selectivity, reducing possible side-effects.[3]

Most peptidomimetics are oligomers derived from natural peptides with modifications in the backbone and/or in the side-chains. In addition, there are also non-oligomeric peptidomimetic compounds that have no apparent structural similarity with the original peptide. These are usually discovered by random screening of natural products and compound libraries. A classical example is morphine (**1**) that binds an opioid receptor, mimicking the peptide β-endorphin (composed of 31 amino acids), and has long been used as an analgesic (Figure 1).[2] Rational design and computer modeling have also led to the development of therapeutically relevant peptidomimetics like the HIV-protease inhibitor Ritonavir (**2**) (Figure 1).[4-5]

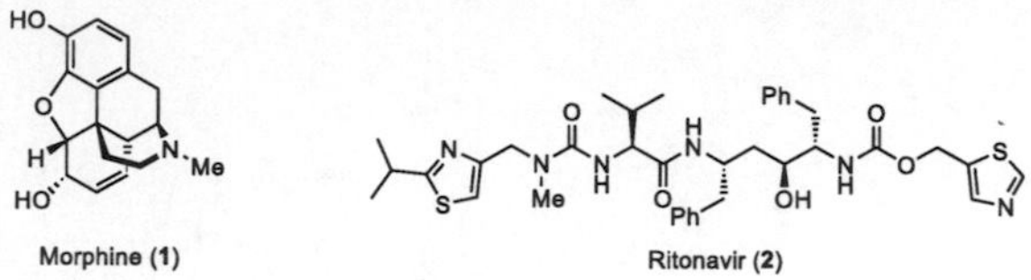

Morphine (1) Ritonavir (2)

Figure 1: Examples of non-oligomeric peptidomimetic drugs.

A commonly used strategy for side-chain modification is the replacement of natural by unnatural amino-acids. For example, switching phenylalanine for the more sterically demanding diphenylalanine led to a potent angiotensin II agonist.[6]

Modification of the peptide backbone has afforded many types of functional peptidomimetics. Figure 2 depicts the most representative examples conserving the amide linkage.

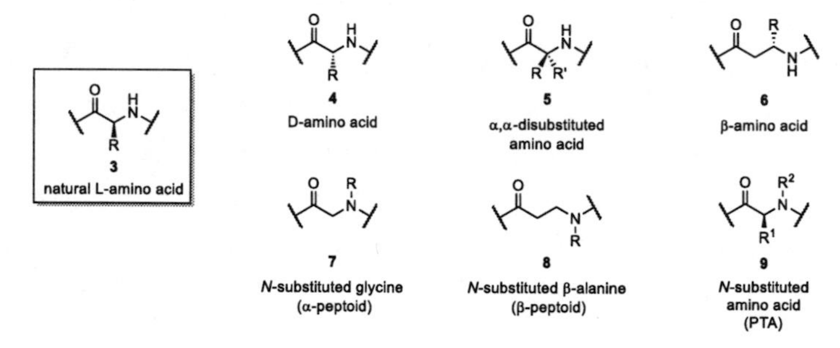

Figure 2: L-Amino acid (**3**) and derived monomer units with modified backbones.

Peptides assembled from D-amino acids (**4**) exhibit a higher proteolytic stability than their natural analogs and have shown promising results in immunological applications.[7] The desire to create peptide analogs with reduced flexibility led to the development of α,α-disubstituted amino acids (**5**). Thanks to the extra substitution at the α-carbon, these oligomers were able to adopt new secondary structures.[8] β-Peptides (**6**), first described by Seebach and coworkers,[9] differ from α-peptides in the presence of an extra methylene unit in their backbone. Despite their increased flexibility, β-peptides are able to form stable secondary structures and a number of biological activities have been reported. They hold special promise in the treatment of autoimmune diseases.[10] The formal shift of the side-chain from the α-carbon to the amide nitrogen afforded new types of peptidomimetics: α- and β-peptoids (**7** and **8**). α-Peptoids have attracted a great amount of interest, probably due to their easily accessible synthesis. Key aspects of α-peptoid research have been summarized in the following chapters. β-peptoids have been less investigated but several applications as antimicrobial agents have been reported.[11] Recently, Gao and Kodadek described the synthesis of a combinatorial library of peptide tertiary amides (PTAs) (**9**).[12] These conformationally constrained oligomers show great potential as protein ligands.

Oligomeric peptidomimetics where the amide bond is replaced by an isostere have also been explored. Some examples are peptidosulfonamides,[13] urea-peptide hybrids,[14] and triazol peptidomimetics.[15-16]

2.2 Peptoids

Peptoids, oligomers of *N*-substituted glycines, are a prominent class of peptidomimetics, pioneered by Bartlett and coworkers in 1992.[17] The word "peptoid" had been previously suggested by Farmer and Ariëns to refer to substances that could mimic the biological activity of a peptide but were structurally different.[18] Bartlett *et al.* specifically applied the term to *N*-substituted glycine oligomers, as it is commonly used today. Peptoids are synthetic regioisomers of peptides in which the typical amino acid side-chain at the α-carbon has been formally shifted to the amide nitrogen (Figure 2). This structural change improves their stability against enzymatic degradation[19] as well as their membrane-permeability,[20] both desirable characteristics for therapeutic applications.

Unlike peptides, peptoids do not contain stereogenic centers in their backbone. This feature simplifies their synthesis avoiding racemization and epimerization problems. Moreover peptoids are unable to form hydrogen bonds along their backbone, which are responsible for the stabilization of peptide secondary structures. There is a considerable amount of ongoing research about the tridimensional structure of peptoids and how it affects their activity. A few strategies have been developed to favor specific structures and will be discussed in Chapter 2.2.2.

The main advantage of peptoids with respect to other types of peptidomimetics is their ease of synthesis on solid support *via* the submonomer method developed by Zuckermann and coworkers.[21] This modular approach allows the incorporation of a wide diversity of functionalities and can be applied to the synthesis of libraries and high-throughput screening methods.

2.2.1 Peptoid synthesis

Peptoids are usually synthesized on solid-phase, a strategy developed by Merrifield in 1963 for the synthesis of peptides.[22] Here, the first building block (in this case the monomer at the *C*-terminus) is covalently attached to a solid support (typically a cross-linked polystyrene resin) by a linker that can be selectively cleaved at the end of the synthesis. The desired molecule is assembled by a series of heterogeneous reactions alternated with washing steps. Reagents can be used in excess, which may lead to higher yields, and are easily removed by filtration at the end of each reaction, together with unbound by-products. Thanks to the simplicity of operation solid-phase synthesis (SPS) can be easily automated.[23] After cleavage from the resin the final products are usually purified by HPLC (High Performance Liquid Chromatography).

Two common acid-labile linkers are depicted in Figure 3. After acidic cleavage the Rink-amide linker leaves an amide group at the *C*-terminus, whereas the chlorotrityl chloride linker leaves a carboxylic acid, which can be further modified e.g. *via* cyclization or bioconjugation. This second linker can also be cleaved under mild conditions (1,1,1,3,3,3-hexafluoroisopropanol, HFIP) avoiding removal of acid-labile protecting groups.

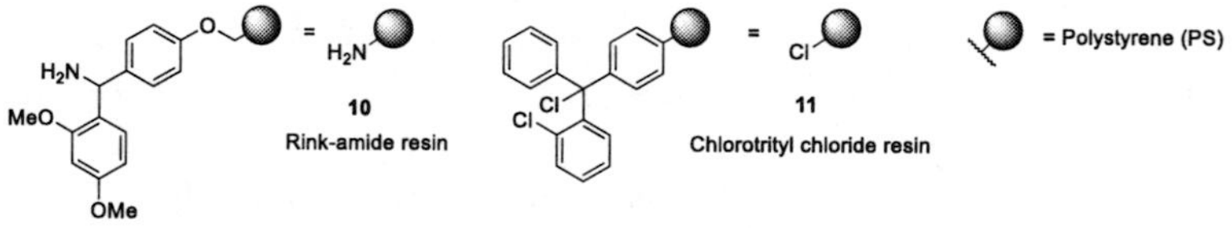

H_2N MeO OMe = H_2N **10** Rink-amide resin Cl Cl = Cl **11** Chlorotrityl chloride resin = Polystyrene (PS)

Figure 3: Chemical structures of the Rink-amide- (**10**) and chlorotrityl- (**11**) functionalized resins for solid-phase synthesis.

In addition to the traditional solid-phase synthesis on polystyrene resins, peptides and peptoids have also been synthesized on cellulose membranes (SPOT synthesis).[24-25] This support is particularly suitable for the screening of combinatorial libraries, since several biological tests can be performed directly on the membrane. There are also examples of peptoid synthesis in solution.[26-27]

On solid-phase, peptoid oligomers can be synthesized *via* two main approaches known as monomer and submonomer methods.

Monomer method

The monomer method is analogous to the solid-phase synthesis of peptides. Instead of amino acids, the peptoid building blocks (monomers) are *N*-substituted glycines.[17] These monomers need to be previously synthesized and protected at the *N*-terminus with 9*H*-fluoren-9-ylmethoxycarbonyl (Fmoc). If the side-chains contain additional reactive positions, these should be orthogonally protected. Bartlett and coworkers[17] reported two main routes for the synthesis of the *N*-substituted glycines (**14**) needed as building blocks: reductive amination of the primary amine **12** (where R is the desired side-chain) with glyoxylic acid (**13**) (route A, Scheme 1) and nucleophilic substitution of amine **12** with chloroacetic acid (**16**) (route B, Scheme 1). Similar to the latter, nucleophilic substitution with ethyl bromoacetate (**17**) followed by saponification has also been described (route C, Scheme 1).[28] In addition monomer **21** could be obtained by Michael addition of glycine (**19**) to acrylamide (**20**) (route D, Scheme 1).[17] Final reaction with Fmoc-succinimide affords the protected monomer **15**.

Scheme 1: Synthetic routes for the preparation of Fmoc-protected peptoid monomers (**15**). A) Reductive amination of amine **12** (where R is the desired side-chain) with glyoxylic acid (**13**).[17] B) Nucleophilic substitution of amine **12** with chloroacetic acid (**16**).[17] C) Nucleophilic substitution of amine **12** with ethyl bromoacetate (**17**).[28] D) Synthesis of monomer **21** by Michael addition of glycine (**19**) to acrylamide (**20**).[17]

The Fmoc-protected monomers **15** are attached to the resin by means of a peptide-coupling reagent (typically *N,N'*-diisopropylcarbodiimide = DIC, and 1-hydroxybenzotriazol = HOBt). After deprotection of the *N*-terminus with piperidine, a new monomer can be attached. The coupling and deprotection cycle is repeated until the desired peptoid length is reached (Scheme 2). Microwave heating has been shown to improve coupling yields and shorten reaction times.[29-30]

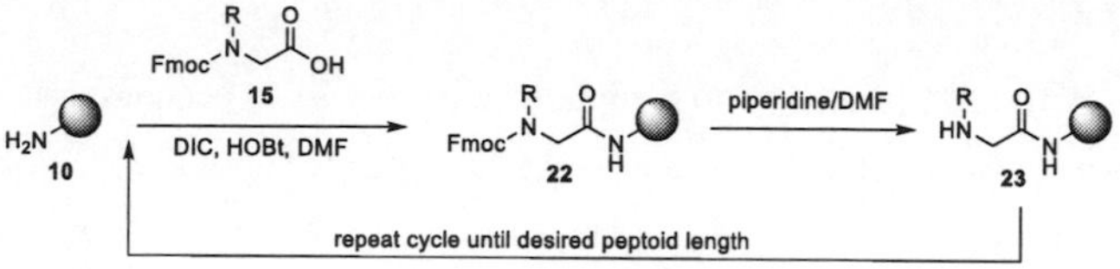

Scheme 2: Peptoid synthesis on solid-phase *via* monomer method.

An advantage of this method is the possibility of monitoring the couplings by quantification *via* UV/VIS spectroscopy of the dibenzofulvene-piperidine adduct generated during Fmoc-deprotection.[28]

However, the extra synthetic effort required for the preparation of the *N*-substituted glycines prior to peptoid assembly, led to the development of the submonomer method.

Submonomer method

In the submonomer approach, developed by Zuckermann and coworkers,[21] each peptoid unit is introduced in two steps: first the coupling of bromoacetic acid activated by DIC, and then a nucleophilic substitution of the bromide by the side-chain as a primary amine (submonomer). The acylation and substitution cycle is repeated until the desired peptoid length is reached (Scheme 3).

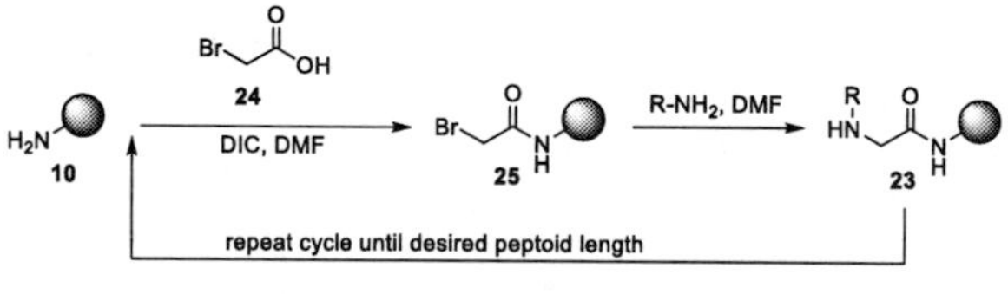

Scheme 3: Peptoid synthesis on solid-phase *via* submonomer method.

The main advantage of this method is that it allows the incorporation of a wide-variety of side-chains as primary amines,[31] many of which are commercially available. Therefore it is particularly suitable for the synthesis of large libraries.[32] A further advantage is that the need of protecting groups is reduced to reactive positions in the side-chains.

While peptoid synthesis *via* the submonomer method is often performed at room temperature, microwave-assisted procedures leading to shorter reaction times have been described as well,[33] especially for electronically deactivated amines.[34]

A variation of the submonomer method employing chloroacetic acid as acylating agent was reported to improve yields and purities in the presence of unprotected nitrogen heterocycles in the side-chains.[35]

After their synthesis peptoids can be further functionalized on solid support *via*, for example, copper-catalyzed alkyne-azide cycloaddition (CuAAC).[36-37] This method can be used for the incorporation of complex or highly functionalized side-chains.

2.2.2 Peptoid structure

The peptoid backbone is more flexible than that of their parent α-peptides. This is due to the lack of substitution at the α-carbon, the inability to stabilize secondary structures *via* backbone hydrogen bonding and the presence of tertiary amides that can readily isomerize between their *cis*- and *trans*-conformations (Scheme 4).

cis ⇌ *trans*

Scheme 4: *cis-trans* Amide equilibrium in peptoids

However, different research groups have developed strategies for the stabilization of peptoid secondary structures, through steric and electronic effects, by incorporation of specific side-chains.

The best characterized peptoid secondary structures are the peptoid helix and the threaded loop (Figure 4).

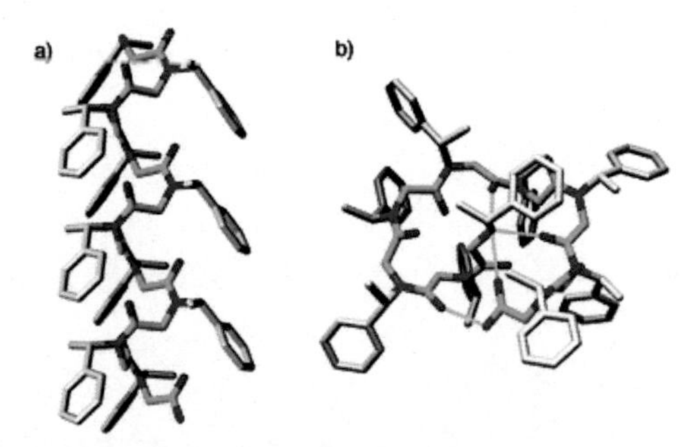

Figure 4: Tridimensional structures of a peptoid helix (a) and loop (b).[38]

The peptoid helix (Figure 4, a) is a polyproline type I helix (PPI) adopted by peptoid oligomers rich in α-chiral side-chains, like (*S*)-1-phenylethyl (**26**) or (*S*)-1-cyclohexylethyl (**27**). Its structure, consisting of three residues per turn and all-*cis* amide bonds, was initially predicted[39] and then supported by circular dichroism (CD) spectroscopy,[40] NMR studies,[41] and X-ray crystallography.[42] Barron and coworkers described a set of sequence requirements for helical stability in peptoids.[43] Helix formation is favored in peptoid oligomers consisting of at least 50% α-chiral residues or presenting an aromatic face parallel to the helix axis, which is achieved by periodic incorporation of aromatic side-chains at *i* and *i*+3 positions. The presence of an α-chiral residue at the *C*-terminus also contributes to the stabilization of the helical structure. Long peptoid oligomers (above 12 residues) tend to adopt more stable helices and are less sensitive to sequence-specific requirements, suggesting cooperative interactions. The most robust peptoid helices described so far were developed by the Blackwell group.[44] These helices, homogeneous in solution, are formed by the bulky (*S*)-1-naphtylethyl side-chains (**28**), which are strong inducers of the *cis*-amide conformation (Figure 5).

The threaded loop (Figure 4, b) is a tridimensional structure exclusive to peptoid nonamers composed by α-chiral residues.[45] It is stabilized by four intramolecular hydrogen bonds and the disruption of these by protic solvents led to disassembly of the loop in favor of the helical structure. The strategic placement of electron-withdrawing aromatic side-chains in peptoid nonamers could be used to favor one structure over the other.[46]

The previous examples show that despite their conformational flexibility, peptoids are able to form defined secondary structures. However the *cis*/*trans* amide isomery hampers the formation of homogeneous structures. In addition, this heterogeneity complicates structure analysis and prediction. Therefore significant efforts have been directed to control the *cis*/*trans* amide equilibrium. Blackwell and coworkers investigated the influence of local non-covalent interactions on the amide-bond conformation of monomeric model systems and demonstrated the importance of local steric and $n \rightarrow \pi^*$ interactions in directing the folding of larger peptoids.[47] Kirshenbaum *et al.* reported the enforcement of *trans*-amide conformation by incorporation of aryl side-chains (**30**).[48] Molecular modeling of *N*-aryl glycine oligomers predicted the formation of helices consisting of all-*trans* amide bonds resembling polyproline type II (PPII) structures. More recently Taillefumier and coworkers developed the triazolium side-chain (**29**), the strongest inducer of *cis*-amide conformation described so far.[49] Moreover this residue has the advantage of maintaining the potential for side-chain diversity.

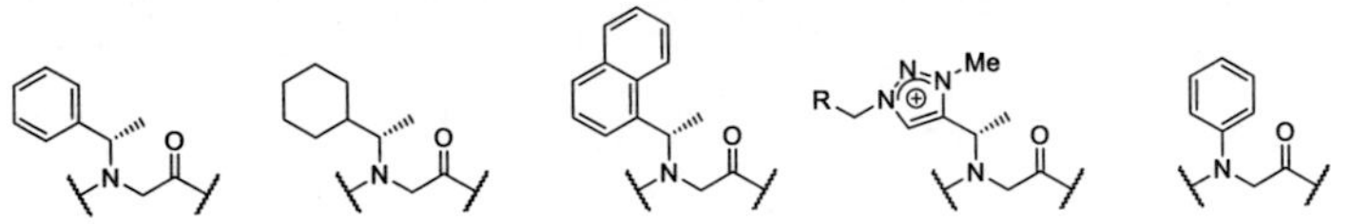

26 27 28 29 30

$K_{cis/trans}$ = 2.0[47] $K_{cis/trans}$ = 1.2[47] $K_{cis/trans}$ = 6.3[44] $K_{cis/trans}$ = 11.7[49] $K_{cis/trans}$ = 0.05[48]

Figure 5: Side-chains used in the stabilization of peptoid secondary structures. $K_{cis/trans}$ calculated in acetonitrile for short monomeric model systems.

These findings allowed the rational design and development of a new peptoid secondary structure, the peptoid ribbon, consisting of alternating *cis* and *trans* amide bonds.[50] This geometry was achieved by combination of (*S*)-1-naphtylethyl (**28**) and aryl (**30**) residues, both of which show defined amide rotamer preferences (Figure 5).

Covalent modifications can also be used to enforce peptoid secondary structure; a prominent example is head-to-tail peptoid cyclization.

2.2.3 Cyclic peptoids

Macrocyclization is a strategy already employed by Nature to endow short oligomeric sequences with defined tridimensional structures. Some examples are non-ribosomal peptides like tyrocidine A and cyclosporine, or the polyketide antibiotic erythromycin.[51] Synthetic macrocyclic peptides and peptidomimetics often exhibit higher biological activities than their corresponding linear analogs.[52-53]

Head-to-tail macrocyclization of peptoids was first reported by Kirshenbaum and coworkers.[54] Different-length peptoid oligomers (up to 20mers) with an even number of residues were successfully cyclized in the presence of PyBOP in a few minutes (Scheme 5).

The conformational constraint introduced by macrocyclization allowed the crystallization of a cyclic peptoid hexamer and octamer. X-Ray analysis of their crystal structures revealed a (*cis-cis-trans*)$_2$ and a (*cis-cis-trans-trans*)$_2$ amide bond pattern for the hexamer and octamer, respectively. Both structures highly resemble peptide β-turns. Thus, peptoid macrocyclization can provide an efficient strategy for the development of turn peptidomimetics.

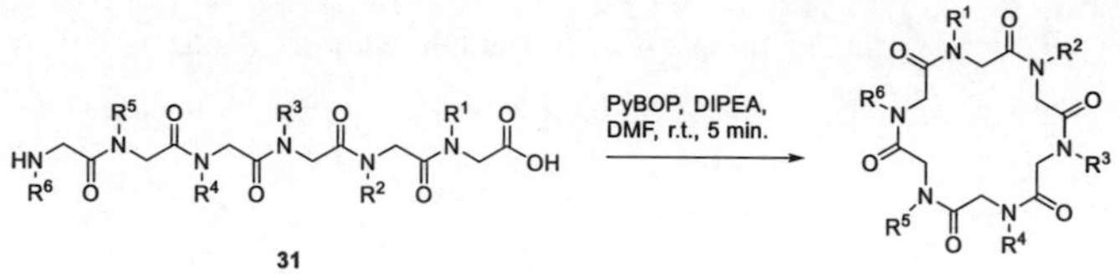

PyBOP, DIPEA, DMF, r.t., 5 min.

31

32

Scheme 5: Head-to-tail peptoid macrocyclization.[54]

Interestingly, in the cyclic hexamer neighboring side-chains were segregated to opposite sides of the macrocycle (Figure 6), suggesting the possibility of designing cyclic peptoids with defined side-chain orientation.

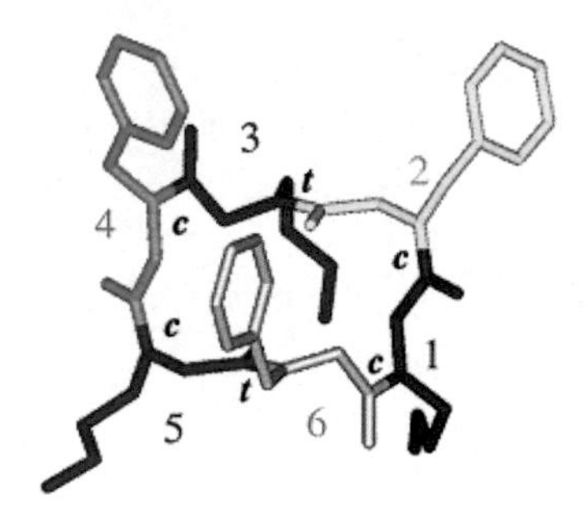

Figure 6: Crystal structure of a cyclic peptoid hexamer.[54] *c* = *cis* amide bond; *t* = *trans* amide bond.

Kirshenbaum and coworkers demonstrated that macrocyclization enhanced antimicrobial activity in a series of peptoids.[55] Peptoid macrocycles are also promising candidates for the development of supramolecular structures. For example, a cyclic octamer crystalized in a tubular structure, that was able to reversibly capture water undergoing a single-crystal to single-crystal transformation.[56] A cyclic hexamer coordinating sodium ions, was used for the generation of a 1D metal organic framework (MOF).[57]

In addition to head-to-tail cyclization, peptoid macrocycles can be built by side-chain ligation. For example, lactam bridges[58] and CuAAC[59] of residues at *i* and *i*+3 positions were used to stabilize peptoid helices. Other strategies to introduce intramolecular linkages in peptoids include olefin metathesis[60] and boronate ester formation.[61] In a different approach, cyclic peptoids have also been obtained *via* *N*-heterocyclic carbene (NHC)-mediated ring-opening polymerization.[62] Additionally constrained bicyclic peptoids could be prepared by bridging an octameric macrocycle *via* intramolecular CuAAC.[63]

2.2.4 Peptoid applications

Since their discovery in the early 1990s a wide variety of peptoid applications have been reported, mostly in the fields of chemical biology and material sciences.

Biological applications include, among others, antibiotic and antitumor agents, diagnosis, and cellular delivery, and will be described in detail in the next chapter.

In recent years peptoids have emerged as a promising class of bioinspired polymers for nanomaterial applications.[64] Their biocompatibility, increased stability,[19] and thermal processability,[65] combined with synthetic diversity and the possibility of precisely tuning their chemical and physical properties,[66] make them attractive candidates for the development of functional biomaterials. Zuckermann and coworkers reported the self-assembly of peptoid oligomers forming ordered nanosheets.[67] The self-assembling peptoid sequences were composed of alternating aromatic hydrophobic residues and either positively or negatively charged side-chains. The assembly mechanism, sequence requirements, and stability of these bilayers have also been investigated.[68-69] The same group described the assembly of a diblock copolymer into a superhelical structure.[70]

In addition, peptoid polymers can act as collagen[71] and polyethylene glycol mimetics.[72] The latter have been successfully applied as antifouling coatings.

Noteworthy applications in other fields include controlling the mineralization of calcium carbonate, as potential CO_2 trap[73] and enantioselective catalysis.[74]

2.2.5 Biologically active peptoids

The numerous biological activities of peptoids can be classified as follows:[38] peptoids that mimic biologically active peptides, peptoids that act as protein ligands or inhibit protein-protein interactions, peptoids for cellular delivery, and peptoids that imitate the tridimensional structure of proteins.

Within the first group, antimicrobial peptoids have gathered a considerable amount of research.[75-77] These peptoids mimic the natural antimicrobial peptides (AMPs) that protect many organisms against bacterial infections. Overall cationic charge and moderate hydrophobicity were found to be important features for antibiotic activity in peptoids.[78] In addition, cyclic peptoids exhibited higher antimicrobial activities than their linear analogs.[55,79]

Peptoid mimics of lung surfactant proteins[80-81] and amylin[82] have also been reported. The first are an example of an application where the secondary structure played an important role in peptoid function. In contrast, increased peptoid flexibility was key for the inhibition of amylin aggregation by the latter.

The screening of combinatorial peptoid libraries has led to the identification of numerous protein ligands.[83-85] The first examples were peptoid ligands of G-protein-coupled receptors reported by Zuckermann and coworkers.[86] This combinatorial strategy was applied by Kodadek *et al.* to identify peptoids binding diagnostically useful antibodies, establishing a general method for the discovery of potential disease biomarkers.[87]

Rational peptidomimetic design has also led to the discovery of new protein ligands. Following this approach, Novartis synthesized a peptoid that selectively binds amyloidogenic misfolded proteins.[88] This finding was applied to the development of a method for quantitative misfolded protein detection, which can be used for the diagnosis of diseases associated with protein misfolding. By mimicking the tridimensional arrangement of the side-chains involved in protein-protein interactions, Appella and coworkers designed inhibitors of the HDM2-p53 complex, associated with tumor aggressiveness and drug resistance.[89] Surprisingly, peptoids lacking a defined secondary structure exhibited higher binding affinities, in this case. Multivalent ligands, where the peptoid moiety is either the active ligand[90] or the scaffold for ligand display,[91] have been used to successfully target relevant receptors in cancer therapy.

The ability of peptoids to act as cellular-delivery agents has been extensively investigated. They were developed to improve the proteolytic resistance and pharmacological properties of the well-established cell-penetrating peptides,[92] originally inspired by the cellular permeability of a cationic sequence in the Tat protein from the HIV virus.[93] The first example of peptoids as molecular transporters was reported by Zuckermann and coworkers.[94] They showed that cationic peptoids were able to complexate DNA and transfect it inside the cells. The transfection efficiencies were improved by using lipid-peptoid conjugates (lipitoids).[95] Two years later, Rothbard *et al.* reported an enhanced cellular uptake for peptoids containing guanidine side-chains, in comparison with the corresponding peptide analogs.[96] Our group explored the cellular uptake of fluorophore-labeled peptoids containing amine or guanidine side-chains.[97] The first were transported to the cytosol whereas the latter preferably accumulated in the nucleus. The structure of these peptoids could be elucidated using NMR and computational methods.[98] Cellular uptake in plant cells was also investigated.[99] Recently, our group reported targeted delivery of lipophilic peptoids into the mitochondria.[100] Cell-penetrating peptoids were also successfully applied to the internalization of lanthanide clusters.[101]

Working toward the development of peptoid-based artificial proteins combining function and structure, the research groups of Dill and Zuckermann designed peptoids that are able to fold into multihelical structures.[102] Moreover, they were able to introduce a zinc-binding motif in two-helix bundles, demonstrating that these architectures can reproduce protein structure and function.[103]

2.3 Short Amino Acid Sequences with Antitumor Activity

Short peptide oligomers exhibiting potent biological activities offer a great opportunity for the development of peptidomimetics because their function is less likely to depend on defined secondary structures. The two peptides chosen as model structures for the development of peptoid peptidomimetics in this work are described in the following chapters.

2.3.1 CD44-v6 Peptides

The receptor tyrosine kinase Met controls migration, differentiation and proliferation processes and is overexpressed in various types of cancer.[104] Met activation requires a ligand (hepatocyte growth factor, HGF) and a CD44 isoform containing the sequence of exon v6 in its extracellular domain.[105] v6-Containing isoforms of CD44 are also overexpressed in different types of tumors. The research group of Orian-Rousseau identified by mutational analysis three amino acids within the v6 sequence of CD44 in rats and humans which are essential for ligand-dependent Met activation[106] (Figure 7, left). Short peptide sequences (minimum pentamers) containing these three amino acids (EWQ in rats and RWH in humans) were reported to disrupt the coreceptor function of CD44 preventing Met activation and subsequent cell-migration (Figure 7, right). Stable peptidomimetic analogs of the reported peptides would be attractive candidates for tumor therapy.

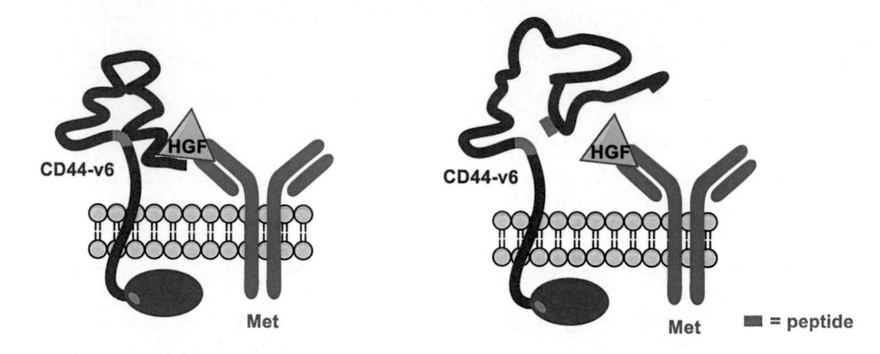

Figure 7: Representation of the Met activation complex formed by the ligand HGF and a v6-containing isoform of CD44 as coreceptor (left). Addition of the CD44-v6 peptide disrupts the coreceptor function and inhibits Met activation (right).

2.3.2 BAG-1 Peptide

Bag-1 (Bcl-2 associated athanogene-1) proteins belong to a family of co-chaperones that possess anti-apoptotic and cell survival properties. Over expression of Bag-1 in cancer cells promotes a number of malignant properties and the level of Bag-1 in prostate cancer correlates with increased tumor aggressiveness.[107] However, in other studies Bag-1 was described as a marker of good prognosis and associated with resistance to chemotherapeutic treatment in prostate cancer.[108] Searching for the molecular basis of this paradox Cato and co-workers identified a region in the Bag-1 protein that inhibits tumor cell growth. Overexpression of this region, strongly inhibited growth of the 22Rv.1 prostate cancer cells in culture and tumor formation in a mouse xenograft tumor model.[109] The effect of this peptide is specific for prostate tumor cells, as it is not observed in benign prostatic hyperplastic cells. The growth inhibitory action of the peptide is associated with its ability to bind glucose regulated proteins (GRP) 75 and 78 that are expressed exclusively on the surface of tumor cells and regulate apoptosis and drug resistance.[110-112]

A sequence of seven amino acids in the peptide (RVMLIGK) was shown to be essential for the binding.[109] Stable peptidomimetics able to reproduce the inhibitory activity of the reported peptide, would be attractive drug candidates for sensitizing cells to chemotherapy.

3. Aims and Objectives

Thanks to their proteolytic resistance, good cell penetrating properties and easily accessible and derivatizable synthesis, peptoids are promising peptidomimetics for therapeutic applications.

The aim of this project was to investigate the ability of peptoids to reproduce the activity of two short amino acid sequences: the CD44-v6 pentamer and the BAG-1 heptamer. These peptides have been shown to inhibit important targets in cancer therapy. Thus, stable peptidomimetics of the reported sequences with improved pharmacokinetic properties would hold high potential as antitumor drugs.

In order to achieve this goal different peptoid oligomers mimicking the model peptides should be designed and synthesized. Therefore, typical solid-phase peptoid synthesis protocols should be adapted to the assembly of the highly functionalized target structures and a robust synthetic procedure allowing for the preparation of a variety of analogs should be established. On a first approach, peptoid analogs with the same side-chains and in the same order as the original peptides should be synthesized. In addition, modified peptoids should be designed aiming to alter functionality, rigidity, or secondary structures. Modifications may include replacement of residues, incorporation of bulky side chains or cyclization. Moreover, a systematic substitution of side-chains, in order to determine their importance for the peptoid's recognition and activity should also be performed. Furthermore, fluorescently-labeled peptoids should be synthesized for visualizing their accumulation in the tumor cells *via* fluorescence microscopy.

Then the biological activity of the synthesized peptoids should be evaluated. Comparison of the inhibitory effect of the peptides and the different peptoid analogs should help to identify essential features for their activity and gain insight into the inhibition mechanism. The optimal candidate should be further investigated *in vivo*.

In addition to mimicking peptide functions, peptidomimetics can also be designed to imitate peptide secondary structures. For that purpose, the ability of peptoids composed of α-chiral amide side-chains to adopt stable secondary structures should be studied.

4. Results and Discussion

4.1 Considerations: Peptoid Nomenclature

In order to simplify the naming of peptoid structures, many groups involved in peptoid research have developed their own nomenclature based on the three-letter amino-acid code used for peptides. However, up to now, there is no universal convention for peptoid side-chain abbreviation. The nomenclature used in this work and described below is inspired by a discussion about the need to develop a unified and systematic nomenclature, which took place at the 8th Peptoid Summit in Berkeley (California) in August 2012.

Here peptoid sequences are named from *N*- to *C*-terminus. An "H" at the beginning of the sequence indicates a free amine at the *N*-terminus, whereas "Ac" denotes that the *N*-terminus is capped with an acyl group. "NH_2" or "OH" at the end of the sequence refer, respectively, to an amide or carboxylic function at the *C*-terminus. The letters SP mean that the peptoid is bound to a solid support. A "c-" preceding the peptoid name indicates a cyclic peptoid.

Each peptoid unit is defined by an alphanumeric four digit code (Figure 8), starting with a capital N, to indicate that the side-chain is attached to the nitrogen. The second digit is a number that refers to the methylene units between the backbone and the first functional group in the side-chain (number 1 can be omitted). The third and fourth digits are an abbreviation of the functional group: am = amine, ad/d = amide, cx = carboxylic acid, gn = guanidine, in = indolyl, im = imidazolyl, mo = methoxy, ph/p = phenyl, sm = methylthio and th = thiol. For example, N5am represents a peptoid residue with 5 methylene units in the side-chain and an amine functionality at the end.

For branched alkyl residues the first number is the length of the longest chain. The third and fourth digits indicate the position and functionality of the branching (m = methyl). For example N32m represents a peptoid residue with a propyl chain that is substituted at position 2 with a methyl. For α-chiral enantiomerically pure residues, the number 1 has been omitted and replaced by the stereo configuration (r or s). The third and fourth digits represent the functional groups attached to the stereocenter. For example Nsmd is a peptoid residue with an *S*-configurated stereocenter at position 1, the substituents of which are a methyl and an amide.

Protecting groups are added as superscript. Glycine (Gly) and sarcosine (Sar) are represented by their usual amino-acid abbreviations.

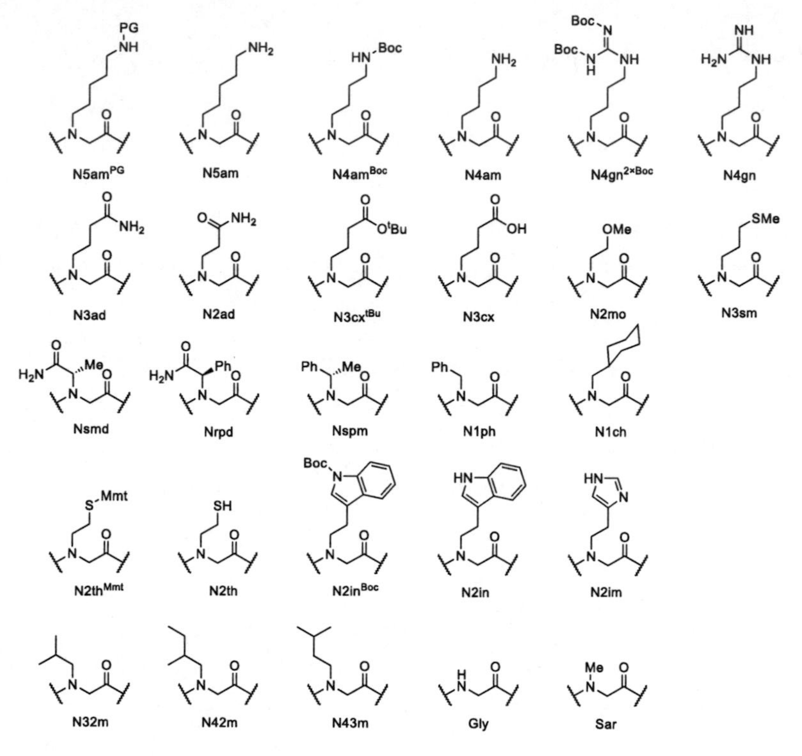

Figure 8: Peptoid side-chain nomenclature.

4.2 CD44-v6 Peptoids

As mentioned in the introduction, the strong activity of short amino acids sequences like the CD44-v6 peptides[106] (pentamers where only the three central residues seem to be essential for inhibition) offers a great opportunity for the development of peptidomimetics.

In order to systematically investigate the ability of peptoids to reproduce the antitumor activity of the CD44-v6 peptides, the first approach was to synthesize peptoids with the same side-chains and in the same order as the active amino acid sequences. Figure 9 shows the reported antitumor CD44-v6 peptides and the peptoid analogs, which were designed as first target structures.

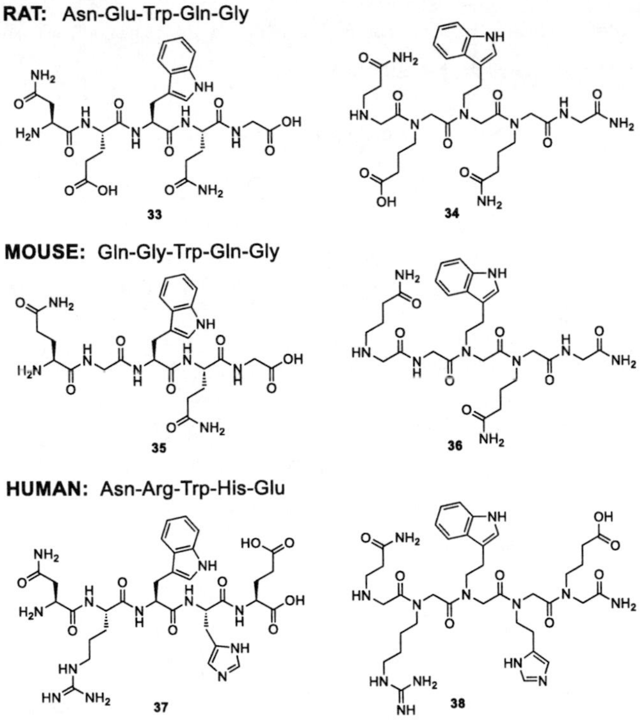

RAT: Asn-Glu-Trp-Gln-Gly

33 34

MOUSE: Gln-Gly-Trp-Gln-Gly

35 36

HUMAN: Asn-Arg-Trp-His-Glu

37 38

Figure 9: Antitumor peptides (left) and their peptoid analogs (right)

The side-chains in the target structures are one methylene unit longer than the original peptide residues. The reason for this design is the advantage of using biogenic amines (obtained from amino acid decarboxylation), many of which are commercially available, as submonomers for peptoid synthesis.

4.2.1 Submonomer synthesis

In peptoid synthesis *via* the submonomer method,[21] the side-chains are introduced as primary amines (submonomers). While many primary amines are commercially available, functionalized side-chains, susceptible of reacting under the coupling conditions, need to be suitably protected. The protecting groups should be orthogonal to Fmoc, in order to be compatible with the monomer method (analog to peptide synthesis)[17] used for the incorporation of glycine or sarcosine units (that cannot be introduced *via* the submonomer method). In general, acid-labile protecting groups are preferred because they will be removed under the acidic conditions required for the cleavage of the peptoid from the Rink-amide resin, avoiding an additional deprotection step.

Figure 10 shows the submonomers needed for the synthesis of target structures **34**, **36** and **38**.

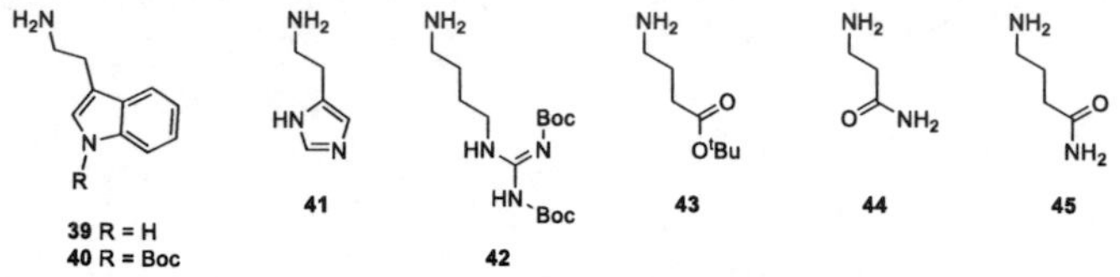

Figure 10: Submonomers used in the synthesis of the target peptoids **34**, **36** and **38**.

Amines **39** and **41** are commercially available and have been previously used unprotected in peptoid synthesis.[35] However, since *N*-heterocycles have also been reported to undergo side-reactions,[113] Boc-protected amine **40** was synthesized as well for comparison. Submonomer **40** could be accessed in three steps starting from tryptamine following a procedure described by Zuckermann *et al.*[114] (Scheme 6).

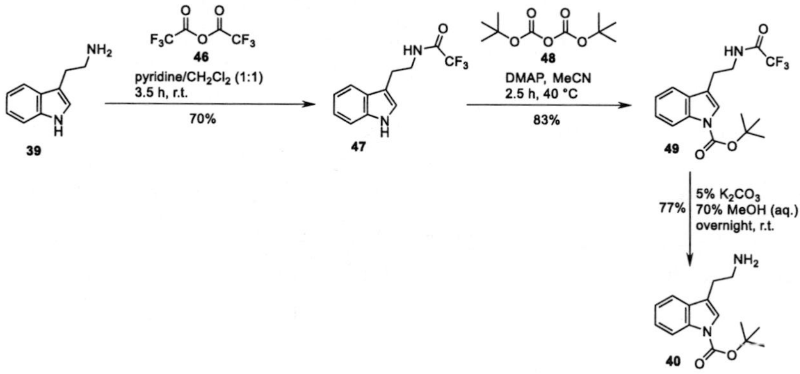

Scheme 6: Synthesis of Boc-protected submonomer **40**.[114]

Boc-protected submonomer **42** was obtained by guanidination of 1,4-diaminobutane (**51**) with *N*,*N'*-di-Boc-1*H*-pyrazole-1-carboxamidin (**50**) (Scheme 7).

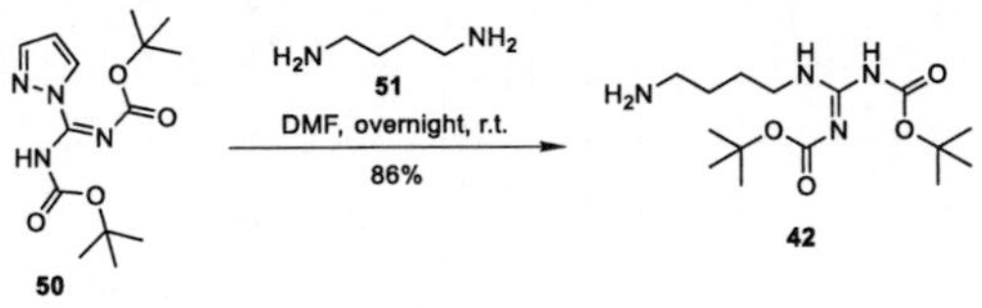

Scheme 7: Synthesis of Boc-protected submonomer **42**.

tert-Butyl protected submonomer **43** was initially synthesized as described by Milić and coworkers.[115] After protection of γ-aminobutyric acid (GABA, **52**) with a Cbz group, the carboxylic function was esterified with *tert*-butanol, *N,N'*-dicyclohexylcarbodiimide (DCC) and catalytic amounts of *N,N'*-dimethylaminopyridine (DMAP) in acceptable yield (Scheme 8, top). Attempts to improve this reaction with *tert*-butyl acetate and perchloric acid failed (Scheme 8, middle), however the yield could be slightly increased *via* Yamaguchi esterification with 2,4,6-trichlorobenzoyl chloride (TCBC)[116] (Scheme 8, bottom). Final Cbz-deprotection afforded **43** in high yield.

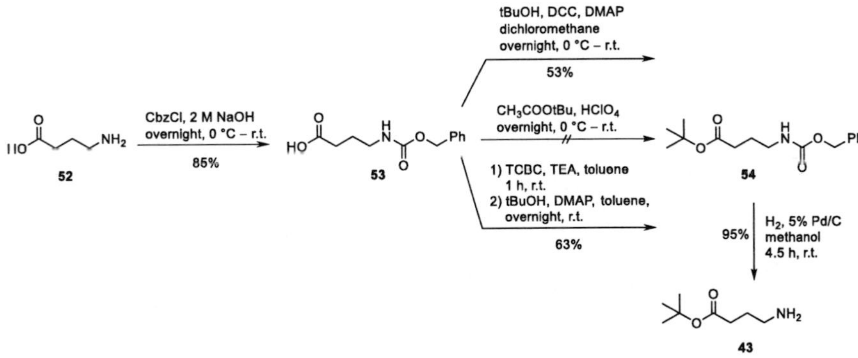

Scheme 8: Synthesis of *tert*-butyl-protected submonomer **43**.

In contrast to the other amines, the synthesis of submonomers **44** and **45** and their incorporation into peptoid sequences was not trivial and will be specifically addressed in the next chapter.

4.2.2 Synthesis of amide-containing submonomers and their incorporation into peptoids

Submonomers mimicking asparagine and glutamine side-chains, like **44** and **45** are rare in peptoid literature and often lack detailed procedures. There are no previous examples of residue N3ad in peptoids. Side-chain N2ad has been coupled as a monomer (Fmoc-protected *N*-alkylated glycine, **Fmoc-21**), which was previously prepared in solution.[17,28] Wenschuh and coworkers reported the use of NaOH in water with 0.05% surfactant (Tween 20) for the incorporation of **44·HCl** as an amine hydrochloride to peptoids on cellulose membranes.[25] The only example of submonomer **44** being introduced without additional reagents was reported by Burgess *et al.*[117]

Initially, the synthesis of submonomers **44** and **45** was planned with Boc as the amine protecting group. Compound **44·HCl** was prepared following a procedure adapted from Keillor and coworkers (Scheme 9).[118] After Boc-protection, β-alanine (**55**) was reacted with *p*-nitrophenyl chloroformate (**57**) in order to convert the carboxylic acid into a good leaving group. Subsequent treatment with ammonia afforded amide **59**. Final Boc-deprotection under acidic conditions led to the desired product as an amine hydrochloride (**44·HCl**). Compound **45·HCl** had been previously prepared in the group following the same route[119] and was readily available.

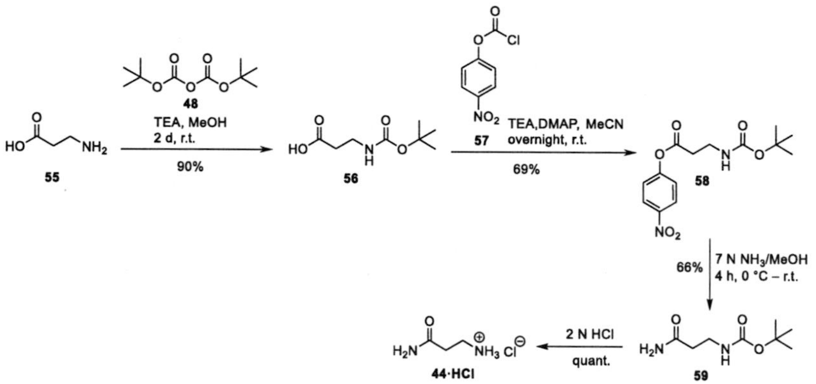

Scheme 9: Synthesis of **44·HCl** following a procedure adapted from Keillor.[118]

The first attempt to incorporate **45·HCl** into a peptoid sequence under the usual submonomer conditions employed for other amines failed (Table 1, entry 1). As shown in Table 1, different bases (DIPEA, K_2CO_3, NaOH) were used to neutralize the hydrochlorides **44·HCl** and **45·HCl** and various solvents (DMF, water) were employed in order to improve the solubility, but in all cases the desired product could not be detected. The method described by Wenschuh and coworkers[25] for the coupling on cellulose membranes was not successful here either

(Table 1, entry 5). The reactions in Table 1 were carried out on different peptoid sequences. The success of the substitution step was determined by mass spectrometry after test cleavage. If the peptoid was further elongated after the failed substitution reaction, the final oligomers were always missing the amide side-chain.

Table 1: Reaction conditions for the incorporation of amine hydrochlorides as submonomers.

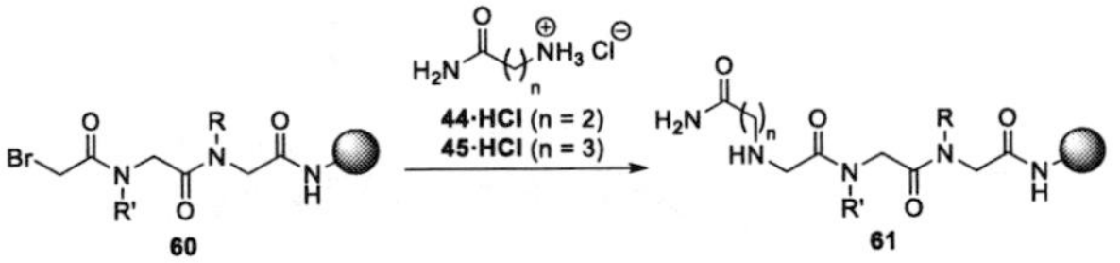

Entry	Submonomer	Base	Conditions	Product[a]
1	**45·HCl**	–	DMF, 60 °C (MW), 15 min	–
2	**45·HCl**	DIPEA (3 equiv.)	DMF, 60 °C (MW), 15 min	–
3	**45·HCl**	DIPEA (3 equiv.)	DMF, r.t., overnight	–
4	**45·HCl**	K_2CO_3 (3 equiv.)	DMF, 60 °C (MW), 15 min	–
5[25]	**44·HCl**	NaOH (1 equiv.)	Water (+Tween), 60 °C (MW), 15 min	–

[a] Presence or absence of the substitution product determined by MALDI-TOF mass spectrometry.

Considering that the low reactivity of the submonomers could be due to a solubility problem, the effect of a lipophilic protecting group on the amide was investigated. Thus, compounds **44·HCl** and **45·HCl** were reacted with trityl alcohol according to Sieber and Riniker (Scheme 10).[120] Unfortunately these reactions did not take place.

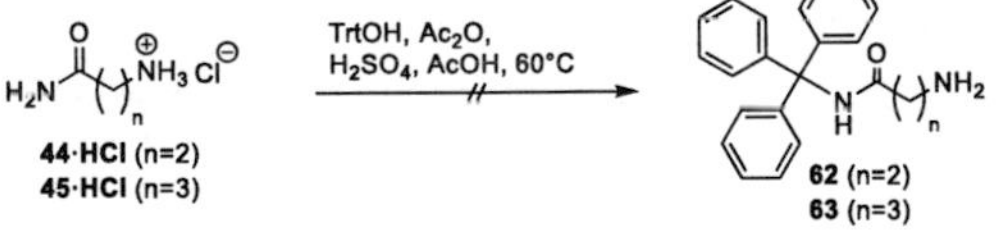

Scheme 10: Attempt for the trityl protection of the amide functionality in compounds **44·HCl** and **45·HCl**.[120]

Finally, the use of Cbz as the amine protecting group in the synthesis of **45** from GABA (**52**) allowed the isolation of the free amine in good overall yield after deprotection *via* hydrogenation (Scheme 11).[121]

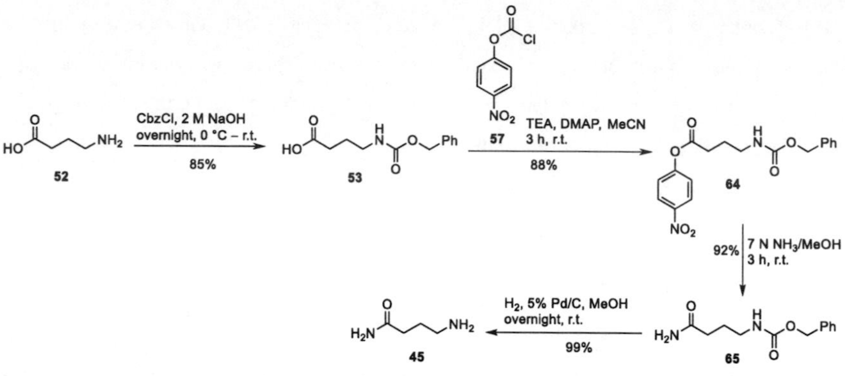

Scheme 11: Synthesis of submonomer **45** as a free base.[121]

Submonomer **44** could also be synthesized as a free base by reduction of cyanoacetamide (**66**). The procedure for nitrile reduction with H_2 over PtO_2 in acetic acid, adapted from Bartlett and coworkers,[122] afforded amine acetate **44·AcOH** that was eluted through a basic anion exchange resin AMBERLITE® IRA400 (OH) to obtain the free amine in very good yield (Scheme 12).

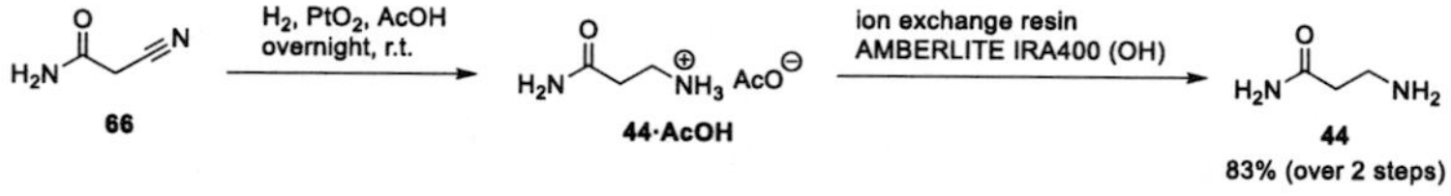

Scheme 12: Synthesis of submonomer **44** as a free base.[121]

Next, the incorporation of the newly synthesized submonomers **44** and **45** in peptoid sequences was investigated.

Reaction conditions for the coupling of **45** were tested on peptoid **68**, which was previously prepared following standard peptoid synthesis protocols on a low-loading Rink-amide resin (Scheme 13).

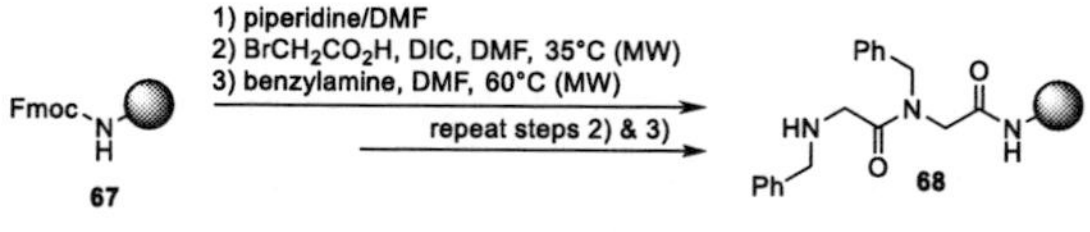

Scheme 13: Synthesis of peptoid **68**.

The different conditions used for the substitution step with **45** are shown in Table 2. The experiments were evaluated by comparison of the HPLC profiles of the crude mixtures. After

1.5 h reaction at room temperature, the product peak represented 31% of the crude mixture. Extension of the reaction time did not generate any changes in the HPLC profile (Table 2, compare entries 1 and 2). No significant difference in purity was observed for microwave-assisted synthesis at 60 °C (Table 2, compare entries 1 and 3). However, these results could be improved by increasing the amount and the polarity of the solvent (switching from DMF to *N*-methyl-2-pyrrolidone (NMP) and diluting the solution to 0.7 M submonomer concentration) (Table 2, entry 4). In this last case a purity of 51% was measured by HPLC.

Table 2: Screening of reaction conditions for the incorporation of **45** into peptoid sequences.

Entry	Conditions (step 2)[a]	Product (%) in crude[b]
1	1 M DMF, r.t., 1.5 h	31
2	1 M DMF, r.t., overnight	31
3	1 M DMF, 60 °C (MW), 45 min	33
4	0.7 M NMP, 60 °C (MW), 45 min	51

[a]Conditions = submonomer concentration, solvent, temperature and time. [b]Percentage of product **69** in the crude mixture determined by HPLC (detection at 218 nm).

With the optimized reaction conditions, submonomers **44** and **45** were introduced in a tetrameric test sequence (Scheme 14). Here, benzylamine and 2-methoxyethylamine were chosen as additional side chains because of their high coupling yields. These experiments were meant to establish first, whether the incorporation of the amide submonomers is possible and, after that, if the sequence can be elongated by a further coupling step. Both submonomers **44** and **45** were successfully coupled and peptoids **70** and **71** were isolated in very good yields after cleavage from resin and HPLC purification.[121]

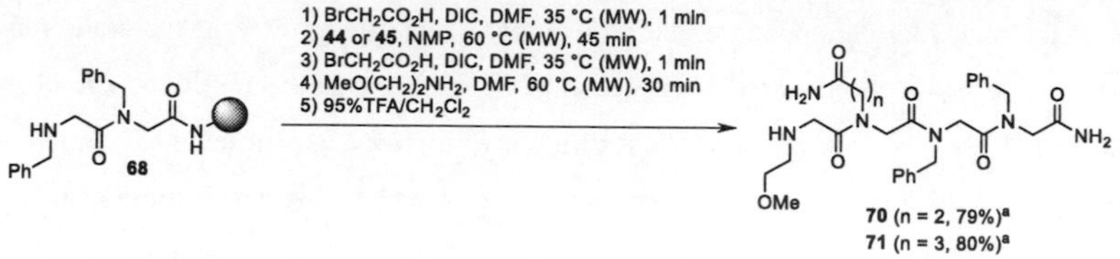

1) $BrCH_2CO_2H$, DIC, DMF, 35 °C (MW), 1 min
2) **44** or **45**, NMP, 60 °C (MW), 45 min
3) $BrCH_2CO_2H$, DIC, DMF, 35 °C (MW), 1 min
4) $MeO(CH_2)_2NH_2$, DMF, 60 °C (MW), 30 min
5) 95%TFA/CH_2Cl_2

68 → **70** (n = 2, 79%)[a], **71** (n = 3, 80%)[a]

Scheme 14: Synthesis of test peptoid sequences containing submonomers **44** or **45**.[a] Yields over 10 steps of the isolated peptoids after HPLC purification.

These findings indicate that the use of the amines as free bases is crucial for the substitution step. The method described here will be exploited in Chapter 4.4 for the investigation of new types of amide-containing peptoids.

During the synthesis of these peptoid test sequences most by-products were identified as the dimers **72**, **73** and **74** resulting from the cross-linking reaction depicted in Scheme 15, where the same amine reacts with two neighboring peptoids during the substitution step.

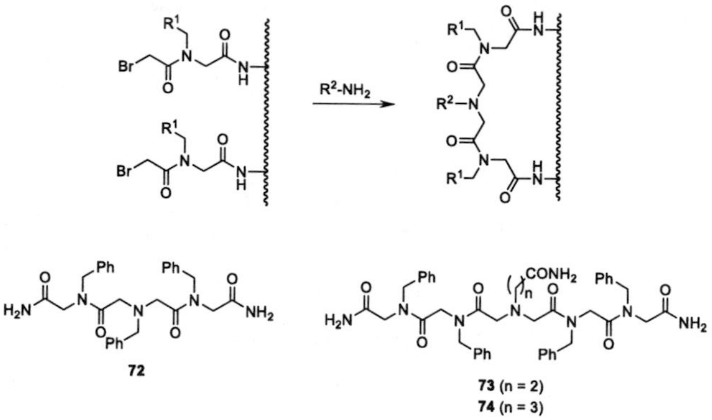

R^2-NH_2

72

73 (n = 2)
74 (n = 3)

Scheme 15: Cross-linking side-reaction leading to dimer formation.

This cross-linking was frequently observed throughout this work and it was found to be highly dependent on the nature of the side-chains (flexibility, directionality, steric hindrance...). For example, the dimer was the only observed product when the bulky α-chiral naphtyl **38** was used as the first submonomer (results not shown).

4.2.3 Optimization of reaction conditions for the synthesis of CD44-v6 peptoids.

At this point, with all the submonomer building blocks in hand, it was possible to address the synthesis of the target structures represented in Figure 9 (pg. 23).

The first syntheses of peptoid **38** *via* the submonomer approach using standard protocols (Scheme 16) resulted in very low purities after cleavage from resin and very difficult and time-consuming HPLC purifications (Figure 11, chromatograms A and B, pg. 35).

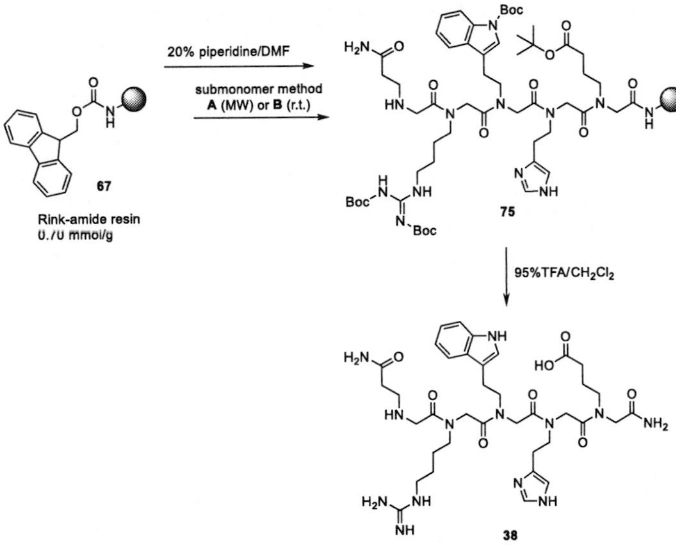

Scheme 16: Synthesis of CD44 v6 human peptoid **38** on high-loading (0.70 mmol/g) Rink-amide resin. Conditions A: *acylation:* $BrCH_2COOH$ (7.9 equiv.) and DIC (7.9 equiv.) 1 M in DMF, 35 °C (MW), 2 min, *substitution*: amine (7.9 equiv.) 1 M in DMF, 60 °C (MW), 15 min: Conditions B: *acylation:* $BrCH_2COOH$ (9 equiv.) 1.2 M in DMF and DIC (7.0 equiv.), r.t., 20 min; *substitution*: amine (8.3 equiv.) 1 M in DMF, r.t., 1–3 h. Substitutions were performed with the following amines: **43**, **41**, **40**, **42** and **44**, in the described order.

As described for the test sequences in Table 2 (see previous chapter), comparison of chromatograms A and B in Figure 11 (pg. 35) shows no significant difference in purity between microwave-assisted synthesis (Scheme 16, conditions A) and room temperature reactions (Scheme 16, conditions B).

Aiming to improve yields and purities, and to facilitate the final isolation by HPLC, the influence of other reaction parameters was investigated (acylating agent, resin loading, capping, post-functionalization in solution). Thus peptoid **80** was synthesized on a low-loading Rink-amide resin (0.39 mmol/g) to avoid cross-linking and interactions between different peptoid oligomers. In addition, bromoacetic acid was replaced by chloroacetic acid as acylating reagent after the incorporation of the first heterocyclic side-chain (Scheme 17). Zuckermann and coworkers had previously reported that the use of chloroacetic acid as acylating agent for sequences with unprotected nitrogen heterocycles afforded purer peptoids.[35] They postulate that the aromatic nitrogen cannot only be temporarily acylated (reversible reaction) but also alkylated (slower but irreversible reaction), and that the reversible acylation leads to accumulation of alkylated side-products. In agreement with their hypothesis they could show that switching from bromo- to chloroacetic acid, the poorer leaving group decreased the rate of the undesired alkylation and increased the purity of the final peptoids.

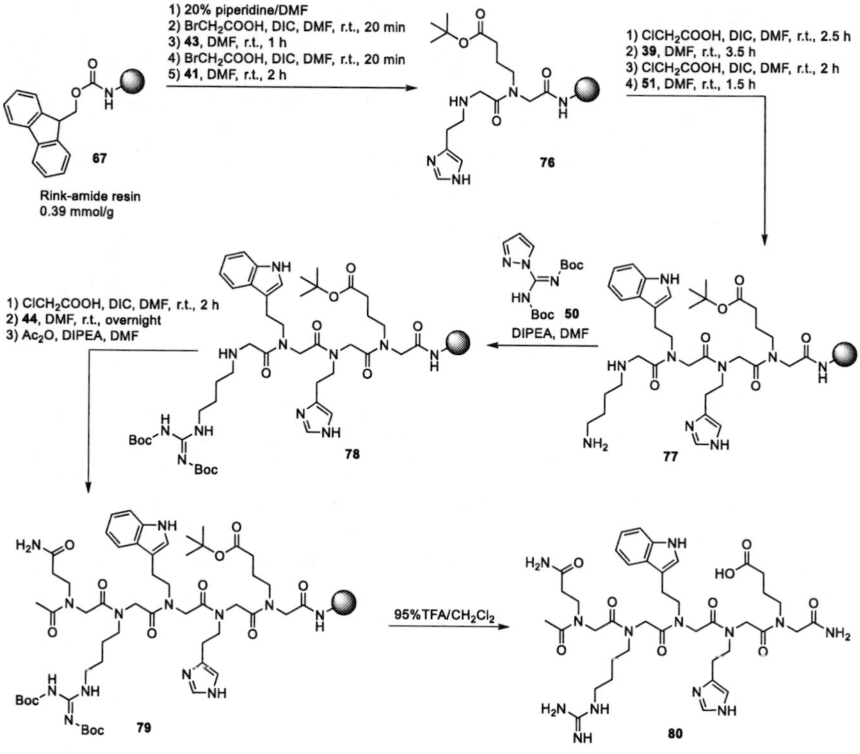

1) 20% piperidine/DMF
2) $BrCH_2COOH$, DIC, DMF, r.t., 20 min
3) **43**, DMF, r.t., 1 h
4) $BrCH_2COOH$, DIC, DMF, r.t., 20 min
5) **41**, DMF, r.t., 2 h

67
Rink-amide resin
0.39 mmol/g

76

1) $ClCH_2COOH$, DIC, DMF, r.t., 2.5 h
2) **39**, DMF, r.t., 3.5 h
3) $ClCH_2COOH$, DIC, DMF, r.t., 2 h
4) **51**, DMF, r.t., 1.5 h

77

50
DIPEA, DMF

78

1) $ClCH_2COOH$, DIC, DMF, r.t., 2 h
2) **44**, DMF, r.t., overnight
3) Ac_2O, DIPEA, DMF

79

95%TFA/CH_2Cl_2

80

Scheme 17: Synthesis of peptoid **80** on low-loading Rink-amide resin with chloroacetic acid as acylating agent and guanidination on solid phase.

Introduction of guanidinated submonomer **42** in the middle of a sequence also seemed to lead to low-purity peptoids. Therefore, in the synthesis of **80** the residue N4am was guanidinated on the solid support to give N4gn. The substitution step was performed with unprotected 1,4-diaminobutane (**51**), which was then directly guanidinated with di-Boc-protected 1*H*-pyrazol-

1-carboxamidine (Scheme 17). Thanks to the low resin loading, no dimers, formed by diaminobutane cross-linkage, were detected by mass spectrometry. Finally, before cleavage from resin, the *N*-terminus of the peptoid was acetylated, to avoid unwanted interactions at this position.

Compared with the previous syntheses, the combination of lower resin loading, the use of chloroacetic acid as acylating agent and the solid-phase guanidination of residue N4am, significantly improved the purity of crude peptoid **80** after cleavage from resin (Figure 11, pg. 35, chromatogram C).

Encouraged by this result, the reaction conditions were further optimized in a new synthesis of **38** (Scheme 18). In this case, the couplings were carried out under microwave irradiation, which, as discussed earlier, does not affect the final purity of the peptoid but leads to shorter reaction times. Like **80,** peptoid **38** was assembled on a low-loading Rink-amide resin and chloroacetic acid was used as acylating agent after residue N2im. However, here submonomer **44** was introduced dissolved in NMP, which already gave better purities in the synthesis of test sequences (see Table 2 in Chapter 4.2.2), and guanidination was performed in solution after cleavage from resin. For this purpose, the side-chain at position 4 from *C*-terminus was incorporated as Boc-protected 1,4-diaminobutane (**81**), which had been previously synthesized as described by Birtalan.[123] After peptoid cleavage from the resin under acidic conditions, the deprotected amine could be guanidinated with 1*H*-pyrazol-1-carboxamidine monohydrochloride without prior purification (Scheme 18). In addition to the cleaner HPLC profile of the crude peptoid (Figure 11, pg. 35, chromatogram D), this route also avoids the need of the expensive Boc-protected guanidinating reagent **50**.

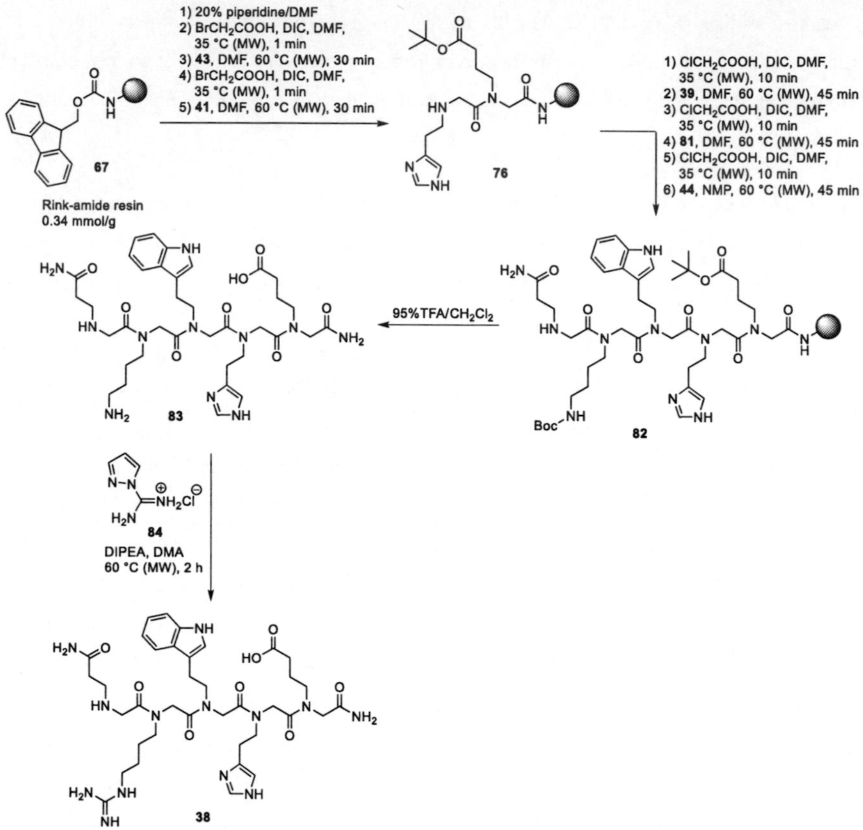

1) 20% piperidine/DMF
2) $BrCH_2COOH$, DIC, DMF, 35 °C (MW), 1 min
3) **43**, DMF, 60 °C (MW), 30 min
4) $BrCH_2COOH$, DIC, DMF, 35 °C (MW), 1 min
5) **41**, DMF, 60 °C (MW), 30 min

67

Rink-amide resin
0.34 mmol/g

76

1) $ClCH_2COOH$, DIC, DMF, 35 °C (MW), 10 min
2) **39**, DMF, 60 °C (MW), 45 min
3) $ClCH_2COOH$, DIC, DMF, 35 °C (MW), 10 min
4) **81**, DMF, 60 °C (MW), 45 min
5) $ClCH_2COOH$, DIC, DMF, 35 °C (MW), 10 min
6) **44**, NMP, 60 °C (MW), 45 min

82

95%TFA/CH_2Cl_2

83

84
DIPEA, DMA
60 °C (MW), 2 h

38

Scheme 18: Microwave-assisted synthesis of **38** on low-loading Rink-amide resin, with chloroacetic acid as acylating agent and post-guanidination in solution.

The bottleneck in the synthesis of highly functionalized peptoids is the final purification *via* HPLC. Therefore, the main criterion for selecting the optimal procedure for the preparation of CD44-v6 peptoids was the ease of isolation of the product peak. Figure 11 shows the section of the chromatogram around the product peak of peptoids **80** and **38** before HPLC purification. Each HPLC profile corresponds to one of the synthetic procedures described above. A clear increase in purity is observed for syntheses on low-loading resin with chloroacetic acid as acylating agent (chromatograms C and D) in comparison with the peptoids assembled on higher-loading resins (chromatograms A and B). Chromatogram D shows the cleanest profile and the easiest isolation. Hence the conditions described in Scheme 18 were selected as optimal for the synthesis of CD44-v6 peptoids.

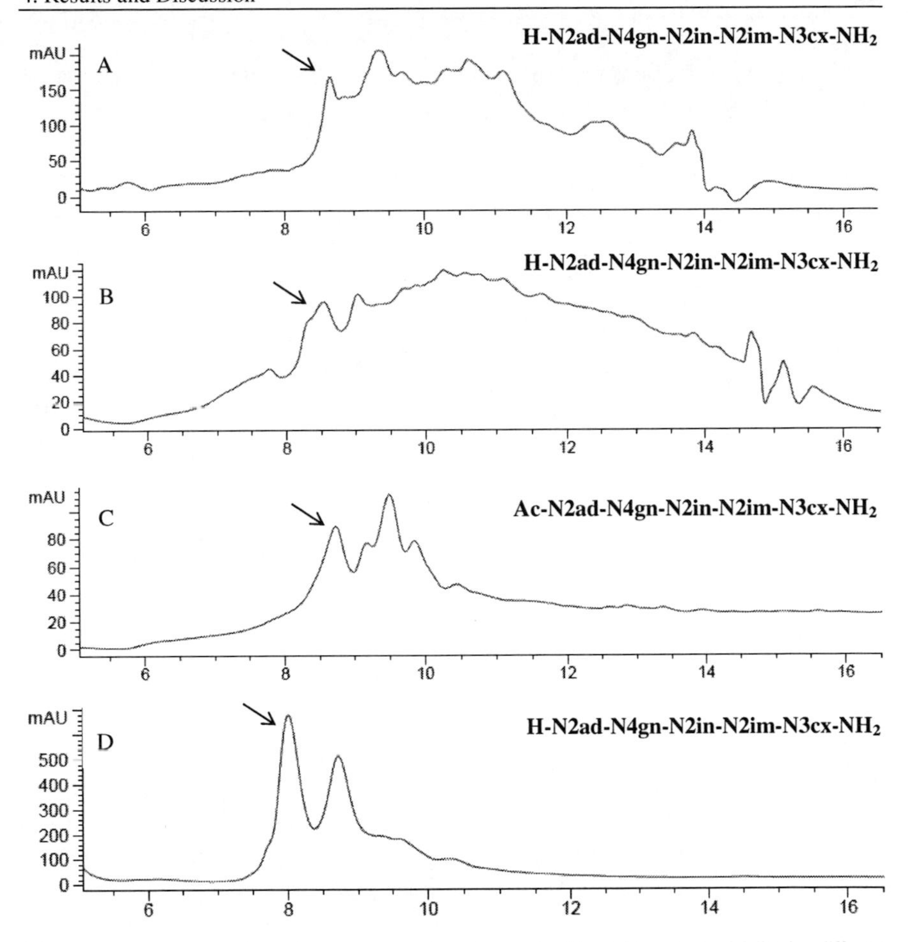

Figure 11: Comparison of chromatogram sections of crude peptoids **38** and **80** synthesized following different procedures. Chromatograms were recorded at 218 nm. HPLC gradient: 5–95% acetonitrile in water + 0.1% TFA over 20 min. The x-axis represents the retention time in minutes. The arrow indicates the product peak. **A**: peptoid **38** synthesized following conditions A in Scheme 16; **B**: peptoid **38** synthesized following conditions B in Scheme 16, **C**: peptoid **80** synthesized as described in Scheme 17, **D:** peptoid **38** synthesized as described in Scheme 18.

Peptoids **Ac-34** and **Ac-36** mimicking the CD44-v6 peptides specific for mouse and rat, were synthesized on a low-loading resin following the optimized reaction conditions described above (Scheme 19). Glycine units were introduced as Fmoc-protected monomers with HOBt and DIC as coupling agents, according to standard peptide synthesis procedures. The coupling was performed twice to ensure maximum yield. The masses of both peptoids where detected by mass spectrometry. However, isolation of peptoid **Ac-34** by HPLC was not possible and peptoid **Ac-36** was only isolated in 80% HPLC purity.

One of the byproducts often detected in the synthesis of these peptoids is dimer **86**, obtained from the cross-linking reaction described in Scheme 15.

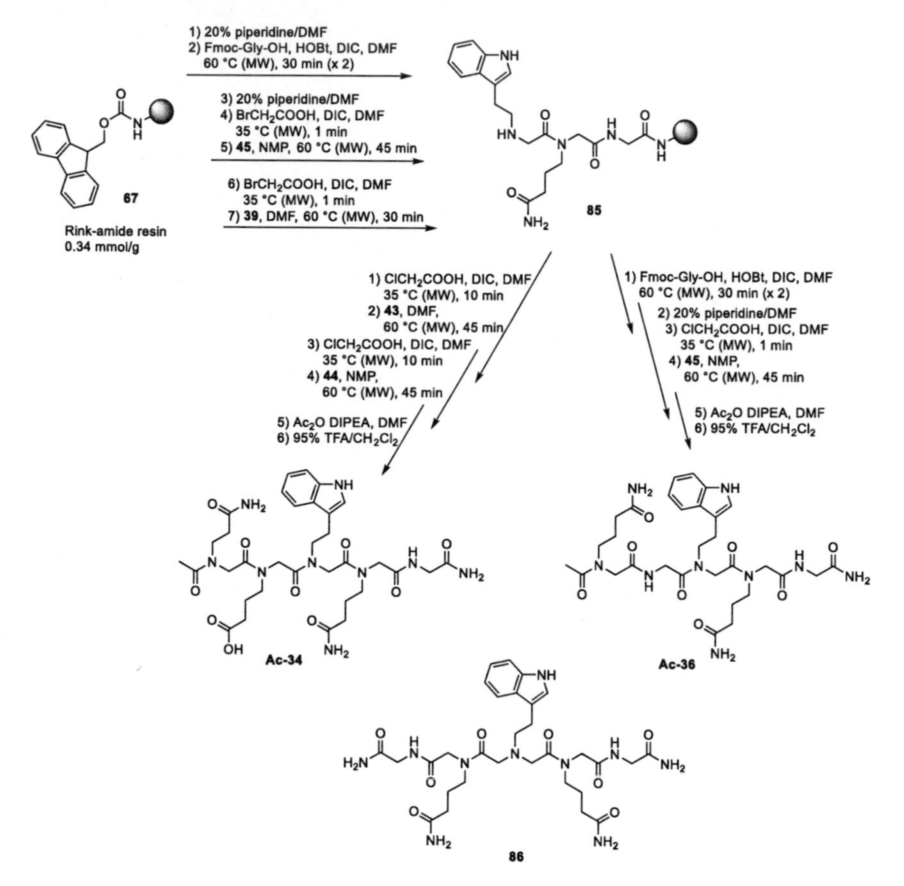

Scheme 19: Microwave-assisted synthesis of peptoids **Ac-34** and **Ac-36** on low-loading resin with chloroacetic acid as acylating agent after incorporation of the first heterocyclic side-chain.

Due to the problematic purifications, peptoids **34** and **36** were not investigated further. Instead the efforts were focused in the synthesis of analogs of the therapeutically more relevant peptoid **38** mimicking the human CD44-v6 peptide.

4.2.4 Synthesis of a peptoid library based on the human CD44-v6 peptide

With the optimized reaction conditions (low-loading resin, chloroacetic acid as acylating agent and final guanidination in solution) a small library of peptoids mimicking the human CD44-v6 peptide **37**, was successfully synthesized and purified (Figure 12).

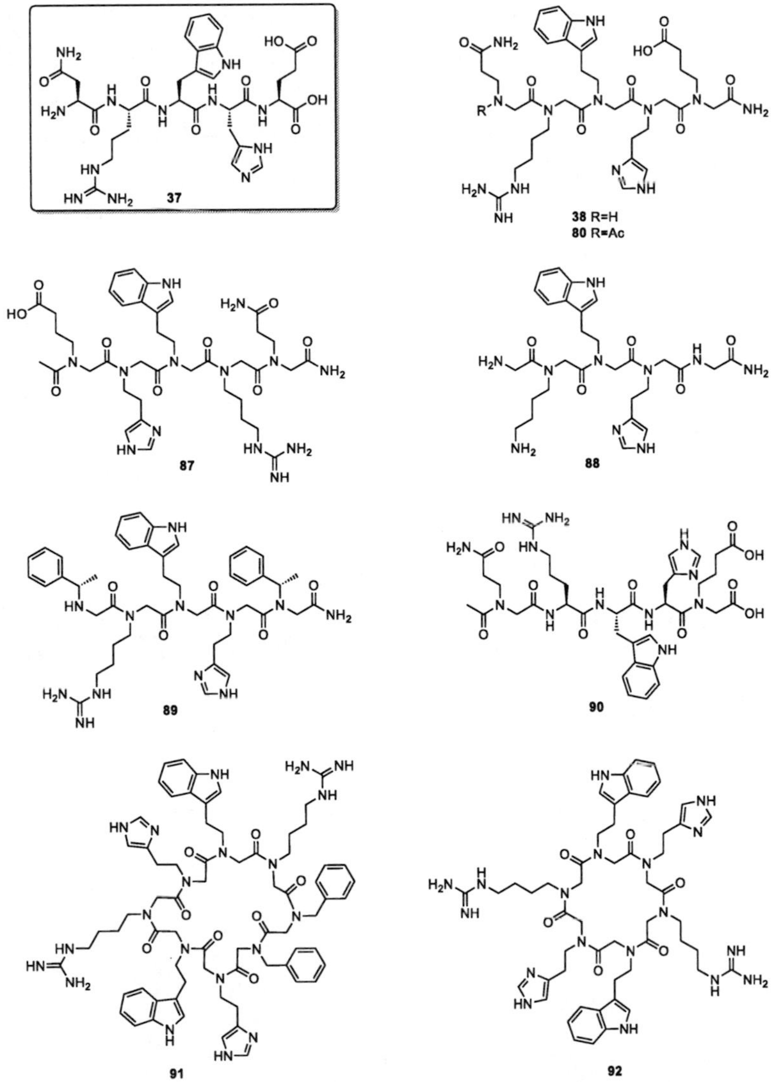

37

38 R=H
80 R=Ac

87

88

89

90

91

92

Figure 12: Synthesized library of peptoids mimicking the human CD44-v6 peptide **37**.

Synthesis of linear peptoids

Peptoids **38** and **80** have the same side-chains and in the same order as the original human antitumor peptide **37**. The only difference between both peptoids is that **80** was acetylated at the *N*-terminus. Their syntheses were already described in Chapter 4.2.2.

Peptoid **87** also has the same side-chains as the human CD44-v6-peptide **37** but the direction of the amide bonds has been reversed (retrosequence). Thus, the relative orientation of the carbonyl groups to the side-chains from the active peptide is preserved. This aspect might be important if the carbonyl group participates in the binding to the receptor, and is probably the reason why, in some cases, retropeptoids show better biological activities than the direct peptide to peptoid translation.[17] Peptoid **87** was synthesized like peptoid **38** in Scheme 18, but the order of introduction of the submonomers was inverted.

Peptoid **88** is a simplified version of the active peptide with fewer functional groups. Here, the end side-chains, that could be exchanged by alanine in the original peptide without affecting its activity,[106] were replaced by glycines. In addition, the guanidine group was substituted by the also protonatable amine function, which reduces the number of synthetic steps. The synthesis, combining the monomer and submonomer methods, is depicted in Scheme 20.

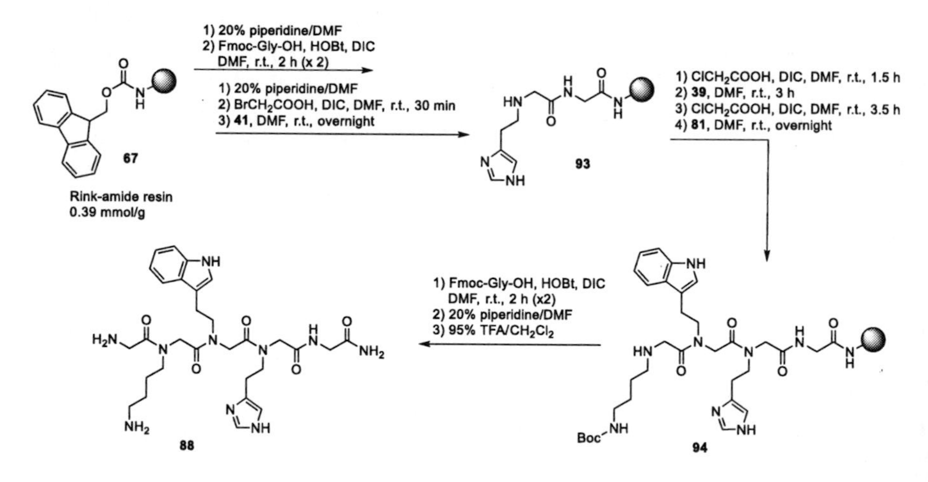

Scheme 20: Synthesis of peptoid **88** on low-loading Rink-amide resin.

As discussed in the introduction, sometimes rigid ligands show increased binding affinities. Therefore, peptoid **89**, with α-chiral side-chains at both ends, and peptide-peptoid hybrid **90**, were designed aiming to reduce peptoid flexibility. Peptoid **89** was synthesized following the procedure represented in Scheme 18, replacing the first and fifth submonomer by commercially available (*S*)-1-phenylethanamine. In peptide-peptoid hybrid **90**, the three central amino acids (shown to be essential for the activity of the peptide in an alanine scan),[106] were conserved, and peptoid residues were introduced at the beginning and at the end of the sequence to increase the proteolytic resistance and improve the pharmacokinetic properties. In the synthesis of **90**, peptoid residues were incorporated *via* the standard submonomer method with bromoacetic acid and DIC for the acylation steps. The central amino acids were coupled through standard peptide synthesis protocols as Fmoc-protected monomers with acid labile protecting groups on their side-chains. HOBt and DIC were used as coupling agents (Scheme 21).

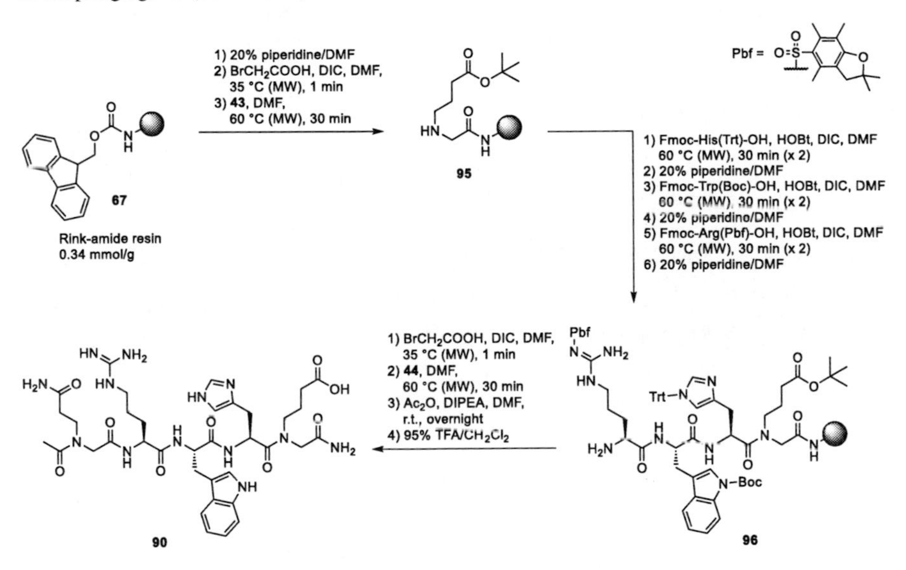

Scheme 21: Synthesis of the peptide-peptoid hybrid **90** on low-loading Rink-amide resin.

Synthesis of cyclic peptoids

A further conformational constraint was introduced in peptoids **91** and **92** *via* cyclization. As described in the introduction (Chapter 2.2.3) Kirshenbaum and coworkers showed that the amide bonds of cyclic peptoid hexamers tend to adopt a *cis-cis-trans-cis-cis-trans* conformation resulting in an alternating 3-up, 3-down orientation of the side-chains.[54] Peptoid **92** was designed based on this idea. In the linear sequence **101**, each of the side-chains responsible for the peptide activity (N2in, N2im and N4gn) is present twice. The order has been chosen assuming that cyclization will drive consecutive residues to opposite faces of the macrocycle. Thus, after cyclization of **101**, identical residues are expected to point to opposite sides of the ring, forming a bidentate-like ligand.

The cyclic octamer **91** was also synthesized for comparison. Benzylamine was chosen as additional submonomer because of its good coupling yields. Here, as a result of the change in size of the macrocycle, a different spatial arrangement of the side-chains was expected.

The linear precursors (**100** and **101**) of cyclic peptoids **91** and **92** were synthesized on chlorotrityl chloride resin as represented in Scheme 22. Cleavage from resin with HFIP left a carboxylic acid at the *C*-terminus. The obtained peptoids **100** and **101** readily underwent head-to-tail cyclization in the presence of PyBOP and DIPEA without prior purification.

Deprotection and guanidination of the amine side-chains of peptoids **102** and **103** were only possible after pre-purification of the crude mixture by HPLC (Scheme 22). Thus, cyclic peptoids **91** and **92**, conformationally constraint analogs of the human CD44-v6 peptide, could be successfully obtained.

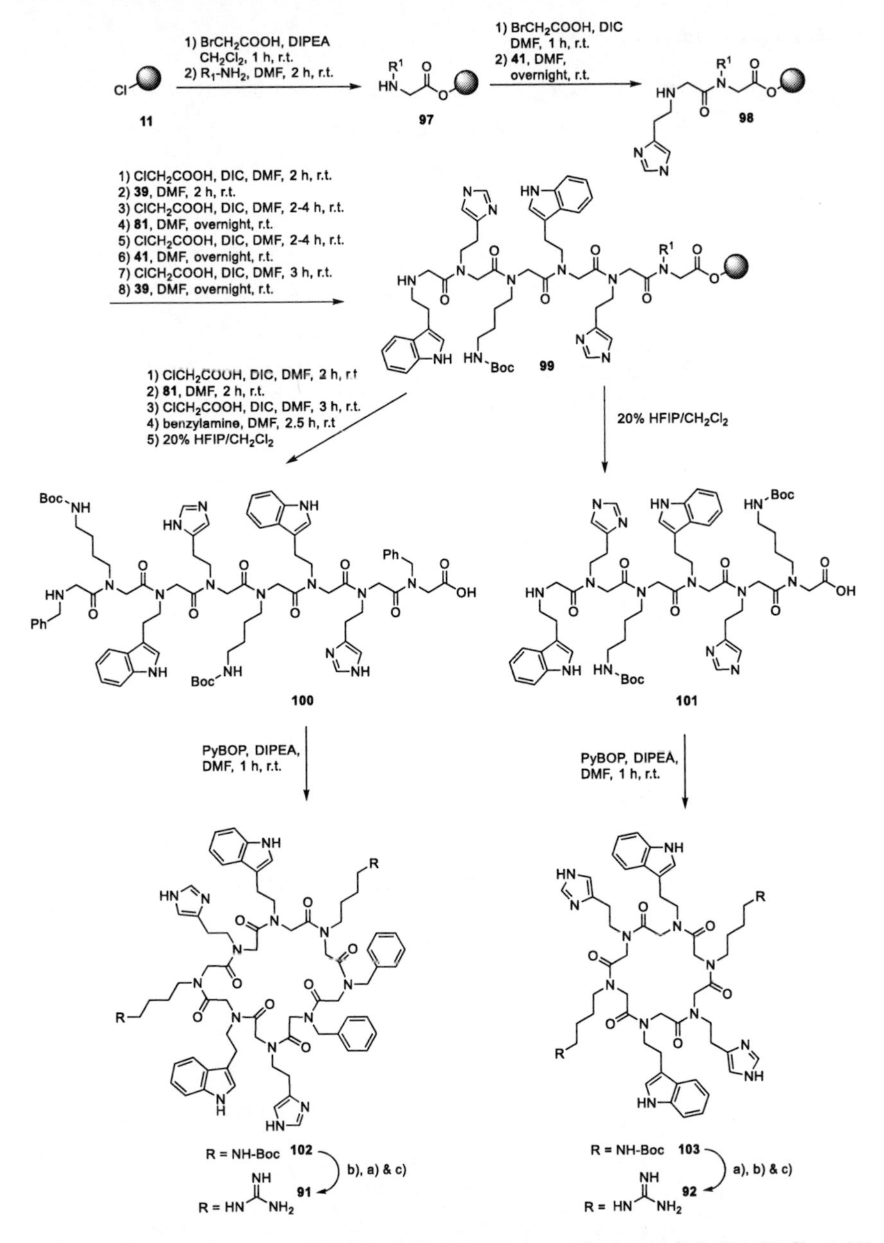

Scheme 22: Synthesis of cyclic peptoids **91** and **92**. a) HPLC pre-purification. b) 50% TFA/CH_2Cl_2. c) 1*H*-pyrazole-1-carboxamidine hydrochloride (**84**), DIPEA, DMA, 60 °C (MW), 2 h.

Synthesis of control peptoids

In addition to the library of analogs of the human CD44-v6 peptide, two unfunctionalized linear and cyclic peptoids (**107** and **110**) were also synthesized as control for the biological experiments. These control peptoids were prepared alternating methyl (Sar) and methoxyethyl (N2mo) side-chains by combining the monomer and submonomer methods (Schemes 23 & 24). Sarcosine was selected as the simplest unfunctionalized representation of a peptoid unit, and the methoxyethyl side-chain was chosen for its good coupling yield and water solubility.

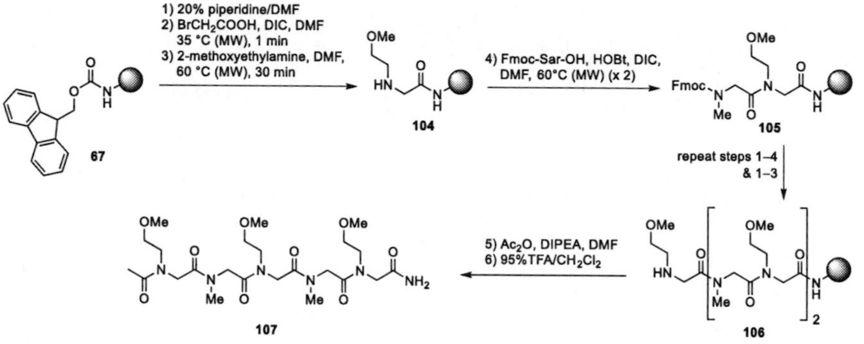

1) 20% piperidine/DMF
2) $BrCH_2COOH$, DIC, DMF 35 °C (MW), 1 min
3) 2-methoxyethylamine, DMF, 60 °C (MW), 30 min
4) Fmoc-Sar-OH, HOBt, DIC, DMF, 60°C (MW) (x 2)
repeat steps 1–4 & 1–3
5) Ac_2O, DIPEA, DMF
6) 95%TFA/CH_2Cl_2
67, 104, 105, 106, 107

Scheme 23: Synthesis of control peptoid **107** by a combination of the monomer and submonomer method.

The linear precursor **109** was synthesized on chlorotrityl chloride resin and then underwent head-to-tail macrocyclization to yield the cyclic control peptoid **110** (Scheme 24).

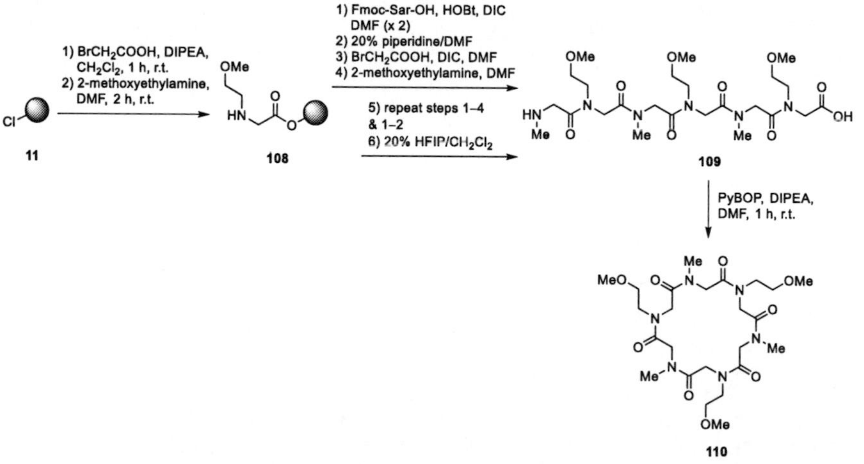

1) $BrCH_2COOH$, DIPEA, CH_2Cl_2, 1 h, r.t.
2) 2-methoxyethylamine, DMF, 2 h, r.t.
1) Fmoc-Sar-OH, HOBt, DIC DMF (x 2)
2) 20% piperidine/DMF
3) $BrCH_2COOH$, DIC, DMF
4) 2-methoxyethylamine, DMF
5) repeat steps 1–4 & 1–2
6) 20% HFIP/CH_2Cl_2
PyBOP, DIPEA, DMF, 1 h, r.t.
11, 108, 109, 110

Scheme 24: Synthesis of cyclic control peptoid **110**.

The yields and purities of the synthesized peptoid library are presented in Table 3. The low yields are due to the difficult HPLC purifications combined with the high purities required for the biological experiments. As expected the highest yields were obtained for peptoids with fewer functional groups (entries 4 and 7).

Table 3: Summary of the synthesized peptoid library.

	Entry	Peptoid	Nr. of steps	Yield (%)	Purity (%)[a]	t_R (min)[b]
Linear	1	**38**	13	3.5	95	7.8
	2	**80**	14	6.0	38-58	8.7-9.0
	3	**87**	14	7.1	>99	8.8
	4	**88**	12	15	97	7.3
	5	**89**	13	1.8	99	10.8
	6	**90**	13	2.5	95	7.9
	7	**107**	14	27	96	10.8
Cyclic	8	**91**	20	1.8	98	11.0
	9	**92**	16	0.5	98	9.1
	10	**110**	14	0.8	96	11.8

[a]Calculated by integration of the HPLC signals at 218 nm. [b]Analytic HPLC (5–95% acetonitrile in water + 0.1% TFA over 20 min).

In conclusion, a library of novel highly functionalized linear and cyclic peptoids, mimicking the antitumor human CD44-v6 peptide was synthesized and purified. The optimized reaction conditions described in Chapter 4.2.2 allowed for the preparation of a variety of analogs. Peptoid **38** could be synthesized in a slightly bigger scale (starting from 700 mg resin) for *in vivo* experiments.

4.2.5 Evaluation of the Met-inhibitory activity of the synthesized peptoid library

The synthesized peptoid library was tested for inhibition of ligand dependent Met activation. The following biological experiments were performed by Mark Schmitt and Dr. Anja von Au in the research group of PD Dr. Véronique Orian-Rousseau.

Met activation requires HGF as a ligand and the v6-containing isoform of CD44 as coreceptor. The activated complex triggers a phosphorylation cascade, which results, among other effects, in the scattering and migration of tumor cells. Based on this mechanism, three different *in vitro* assays were used to detect successful Met inhibition: blocking of

HGF-induced scattering of tumor cells, reduction of the cell migration rate, and decrease of phosphorylated enzymes further down in the cascade triggered by Met (e.g. Erk).

Scattering and migration assays with linear peptoids

Met-inhibition by linear peptoids **38**, **80**, **88** and **89** was first investigated in a scattering assay (Figure 13). Peptoid **107** was used as control.

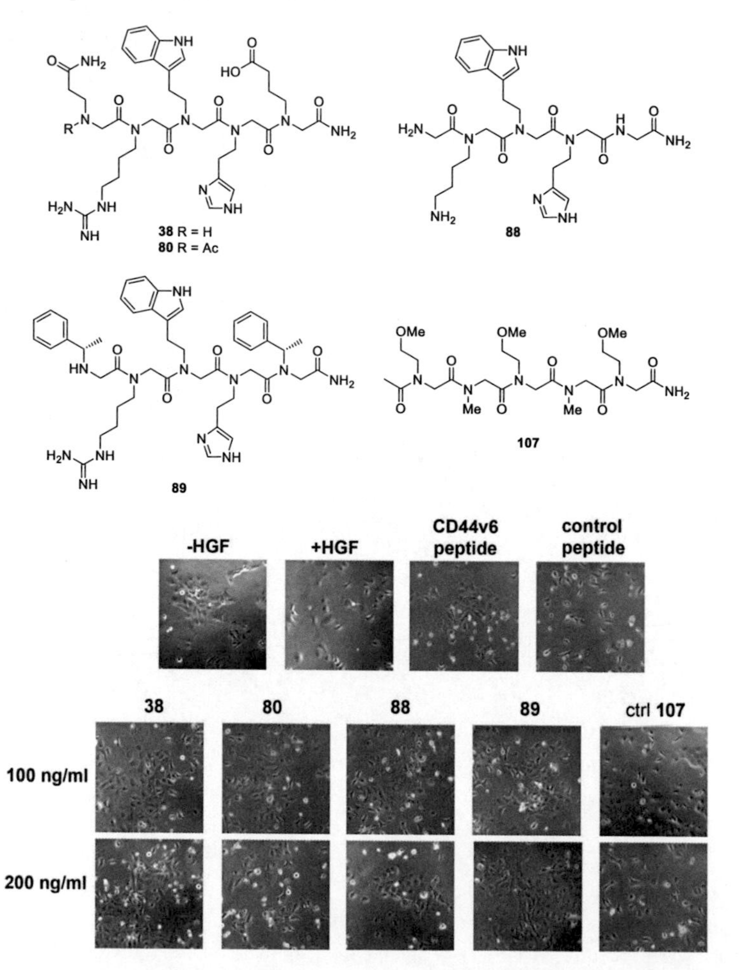

Figure 13: Scattering assay of tongue squamous cell carcinoma SCC25 with peptoids 38, 80, 88 and 89. SCC25 cells were serum starved for 24 hours and incubated with the respective peptides or peptoids for 10 min. Then the cells were induced by HGF (10 ng/mL) for 48 hours. The absence of HGF in one well was used as a negative control and the CD44-v6 peptide (200 ng/mL) as a positive control. HGF induced scattering of SCC25 cells was successfully blocked by peptoids **38**, **80**, **88** and **89**, whereas control peptoid **107** showed no effect.

Induction with HGF caused scattering of tongue squamous carcinoma cells SCC-25 after 24 h. (Figure 13, compare –HGF and + HGF). The scattering was successfully blocked in cells treated with peptoids **38**, **80**, **88** and **89**, as well as the original human CD44-v6 peptide, whereas the control peptoid **107** did not influence the scattering.

It is interesting to note that, in addition to peptoids **38** and **80** (with the same side-chains as the human CD44-v6 peptide), peptoids **88** and **89**, lacking the end side-chains of the original peptide, were still able to inhibit the scattering. For the CD44-v6 peptides it had already been shown that replacement of the end side-chains by alanine did not affect the activity.[106] The blocking effect of the simplified peptoid **88**, is particularly remarkable since it only conserves two of the original peptide functionalities.

As mentioned in the introduction, one of the consequences of Met activation is cell migration. Therefore, the effect of linear peptoids **38**, **80**, **88** and **89** on HGF-induced cell migration was also investigated. For that purpose a "scratch-wound" assay in L3.6pl human pancreatic cancer cells was performed and cell-migration was quantified as percentage of "invaded area" after 24 h (Figure 14) (for the experimental details of the assay see Chapter 5.3).

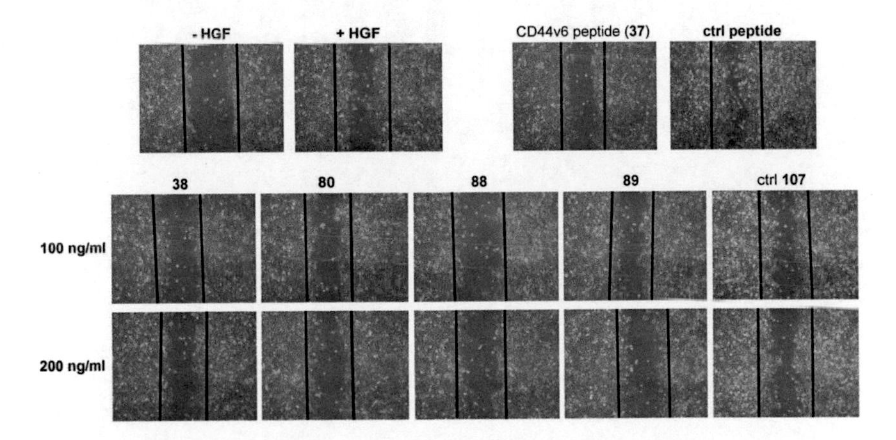

Figure 14: Images of the scratch-wound assay in L3.6pl cells with peptoids 38, 80, 88 and 89. L3.6pl cells were serum starved for 24 h before the assay. In each well a scratch with a pipette tip was performed. The cells were then incubated with the corresponding peptides and peptoids (100 ng/mL or 200 ng/mL) for 10 min at 37 °C before being induced with 10 ng/mL HGF for 24 h. Pictures were taken at time points 0 h and 24 h and the difference in area between the two cell fronts was determined.

The results of the migration assays in L3.6pl cells have been summarized in Chart 1. If Met activation is blocked a decrease in the cell migration rate is expected (the percentage of invaded area should be smaller). This is the effect observed for cells treated with linear peptoids **38**, **80**, **88** and **89** in comparison with control peptoid **107** (Chart 1). Among these, peptoid **38** showed the strongest suppression of HGF induced cell migration in comparison with the control. As observed for the scattering assay, peptoids **88** and **89** (lacking the end side-chains of the original CD44-v6 peptide) also showed inhibitory activity in the migration experiments.

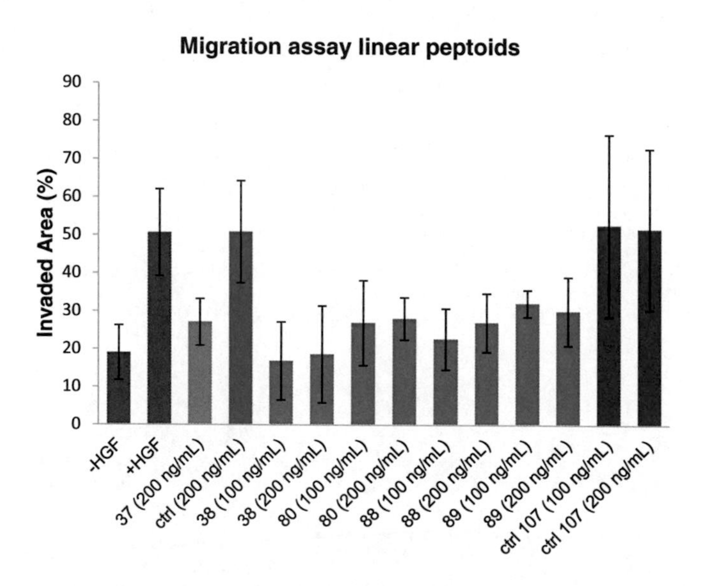

Chart 1: Summarized results of the scratch wound assays with L3.6pl pancreas carcinoma cells. The percentage of invaded area was measured after 24 h (n = 4–5). Purple: controls without peptides/peptoids. Green: CD44-v6 peptide (**37**, positive control) and control peptide. Blue: peptoids.

The ability of peptide-peptoid hybrid **90** and retropeptoid **87** to inhibit HGF-induced cell migration was also investigated (Chart 2). Peptide-peptoid hybrid **90** successfully reduced the migration rate in comparison with the control. However, retropeptoid **87** did not exhibit any inhibitory activity.

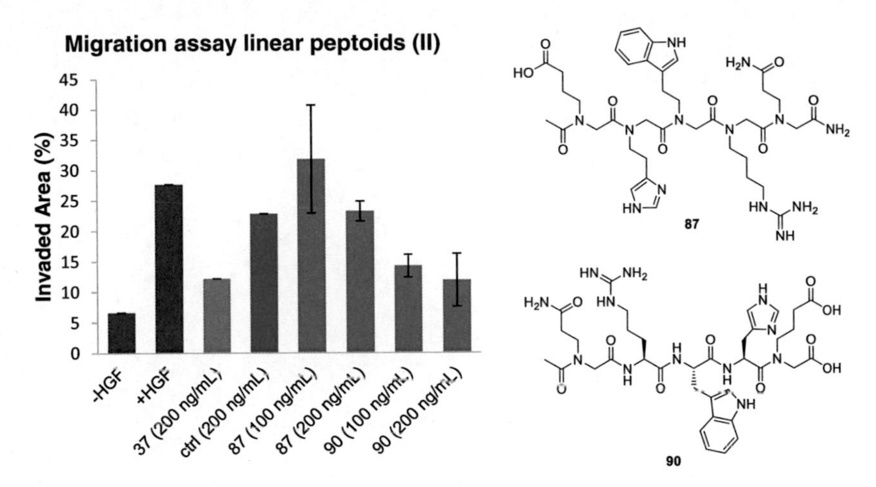

Chart 2: Summarized results of the scratch wound assays with L3.6pl pancreas carcinoma cells for peptoids 87 and 90. The percentage of invaded area was measured after 24 h (n = 3–4). Purple: controls without peptides/peptoids. Green: CD44-v6 peptide (**37**, positive control) and control peptide. Blue: peptoids.

Scattering and Migration experiments with cyclic peptoids

The potential of cyclic peptoids **91** and **92** to inhibit ligand-dependent Met activation was also evaluated in scattering and migration assays.

Both cyclic peptoids **91** and **92** were able to inhibit scattering of SCC25 cells in comparison with the control peptide (Figure 15) whereas the cyclic control **110** showed no effect.

HGF-induced migration of L3.6pl cells was also successfully blocked by both cyclic peptoids **91** and **92** in comparison with the control (Chart 3).

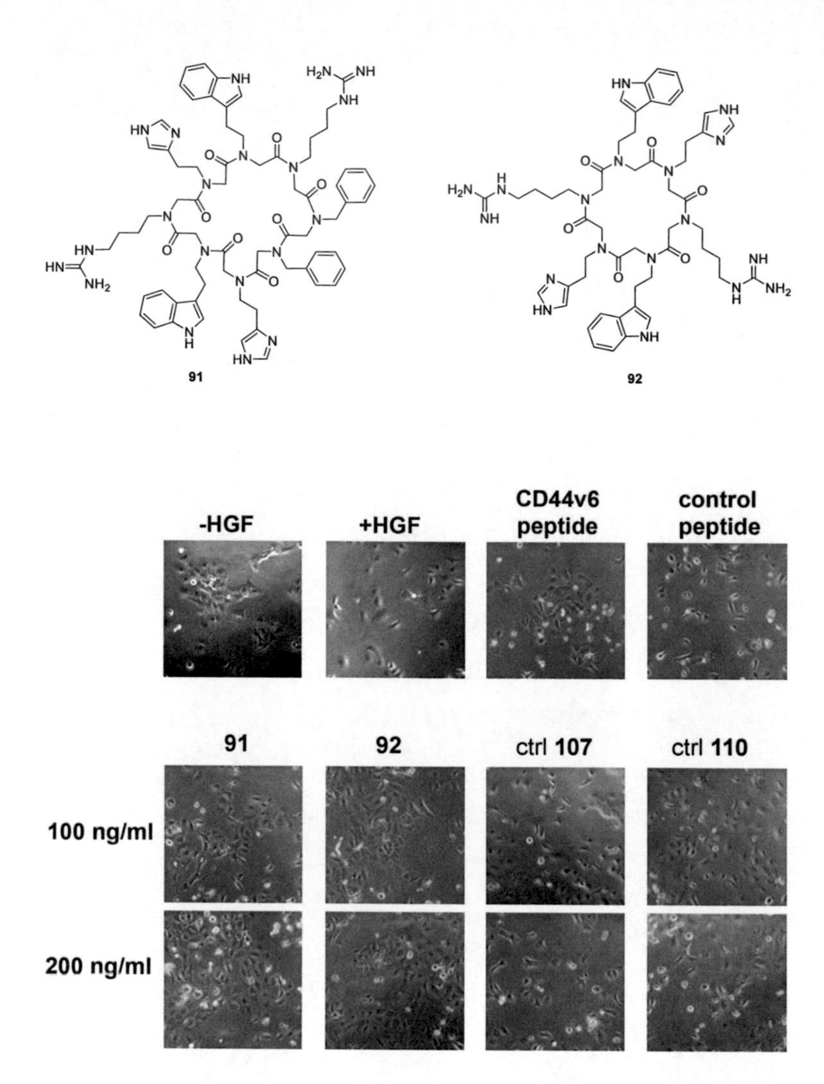

Figure 15: Scattering of tongue squamous cell carcinoma SCC25 assay with cyclic peptoids 91 and 92. SCC25cells were serum starved for 24 hours and incubated with the respective peptides or peptoids for 5 min. Then the cells were induced by HGF (10 ng/mL) for 48 hours. The absence of HGF in one well was used as a negative control and the CD44-v6 peptide **37** (200 ng/mL) as a positive control. HGF induced scattering of SCC25 cells was successfully blocked by both cyclic peptoids **91** and **92**, whereas the cyclic control **110** showed no effect.

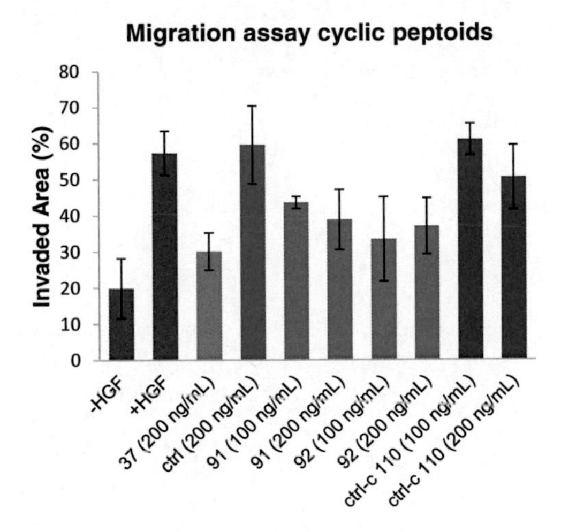

Chart 3: Summarized results of the scratch wound assays with L3.6pl pancreas carcinoma cells for peptoids 91 and 92. The percentage of invaded area was measured after 24 h (n = 5). Purple: controls without peptides/peptoids. Green: CD44-v6 peptide (**37**, positive control) and control peptide. Blue: peptoids.

Altogether these results seem to indicate that the order and functionality of the three central residues is important for the inhibitory activity but conformational flexibility and side-chain orientation do not appear to play an important role. Structure-activity relationships (SAR) will be discussed in detail in the next chapter.

Inhibition of HGF-induced Met signaling

To further confirm the results of the scattering and migration assays, three peptoids representing the functional and structural diversity of the library were tested for inhibition of Met signaling. Thus, the ability of peptoid **38** (with the same side-chains as the original peptide), simplified peptoid **88**, and cyclic peptoid **91** to block HGF-induced Met activation in embryonic kidney cells was evaluated. The cells treated with the peptoids presented reduced levels of phosphorylated Erk (an enzyme further down the phosphorylation cascade triggered by Met), in comparison to those incubated with the control peptide, indicating inhibition of Met signaling (Figure 16). The same effect was observed for the original human CD44-v6 peptide **37**.

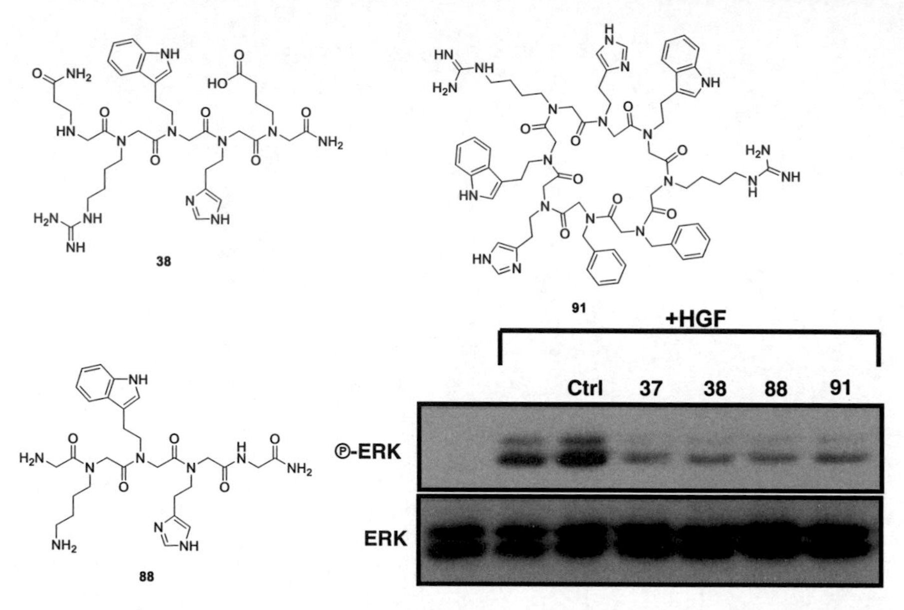

Figure 16: Inhibition of HGF induced Met activation and scattering of human embryonic kidney cells HEK293. HEK293 cells were serum starved for 24 h and incubated with the respective peptides or peptoids (100 ng/mL) for 5 min. Then the cells were induced by HGF (20 ng/mL) for further 5 min. CD44-v6 peptide (**37**) as well as peptoids **38**, **88** and **91** could successfully block Erk phosphorylation and therefore Met signaling in comparison to HGF activated cells alone or the control peptide.

4.2.6 Structure activity relationships in the human CD44-v6 peptoids

Sarcosine scan

In order to investigate which residues are responsible for the inhibitory activity, a "sarcosine scan" (evaluation of the activity of the peptoids resulting from replacement of one side-chain at a time by methyl), equivalent to the alanine scan in peptides, was performed. For that purpose, five new peptoids based on the CD44-v6 peptide (**37**) and exchanging one residue at a time for sarcosine were synthesized and purified (Figure 17). These peptoids are referred to as **Sar-X** where X stands for the position of the residue exchanged for sarcosine (starting from the peptoid carboxy end).

Simultaneously to these investigations, Barron and coworkers reported the use of the "sarcosine scanning" strategy in peptides in order to asses which positions would tolerate peptoid substitution.[124]

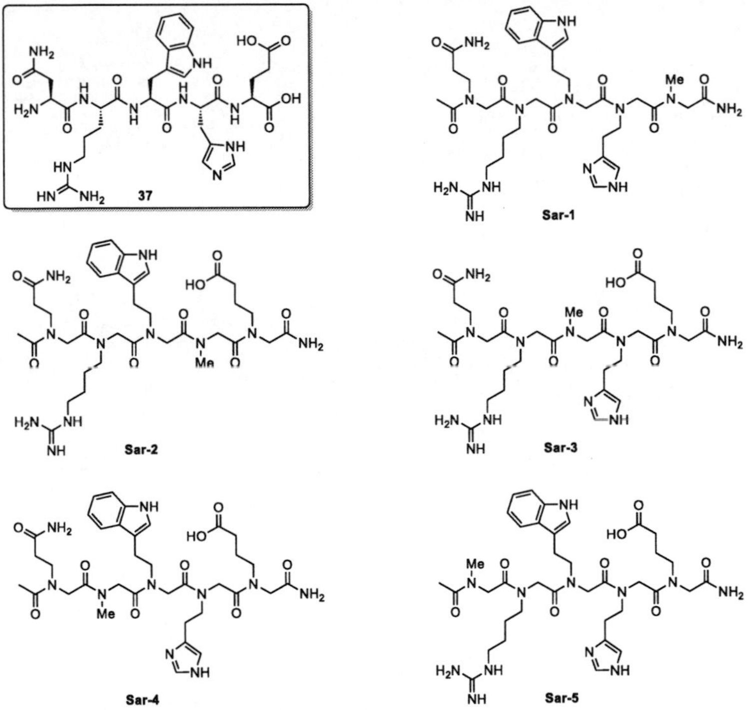

Figure 17: Synthesized peptoids for the sarcosine scan.

The synthesis of the Sar-peptoids was carried out on Rink-amide resin with the optimized conditions described for peptoid **38**. All residues were introduced *via* the submonomer method except sarcosine, which was coupled, at the corresponding position in each peptoid, as an Fmoc-protected monomer with HOBt and DIC as coupling reagents. Before cleavage from the resin, the peptoids were capped at the *N*-terminus. The arginine side-chain was introduced as a protected amine and guanidinated in solution after resin cleavage and pre-purification in HPLC. As an illustrative example synthesis of peptoid **Sar-1** is depicted in Scheme 25. The other Sar-peptoids were synthesized in an identical manner, only differing in the position of the sarcosine unit.

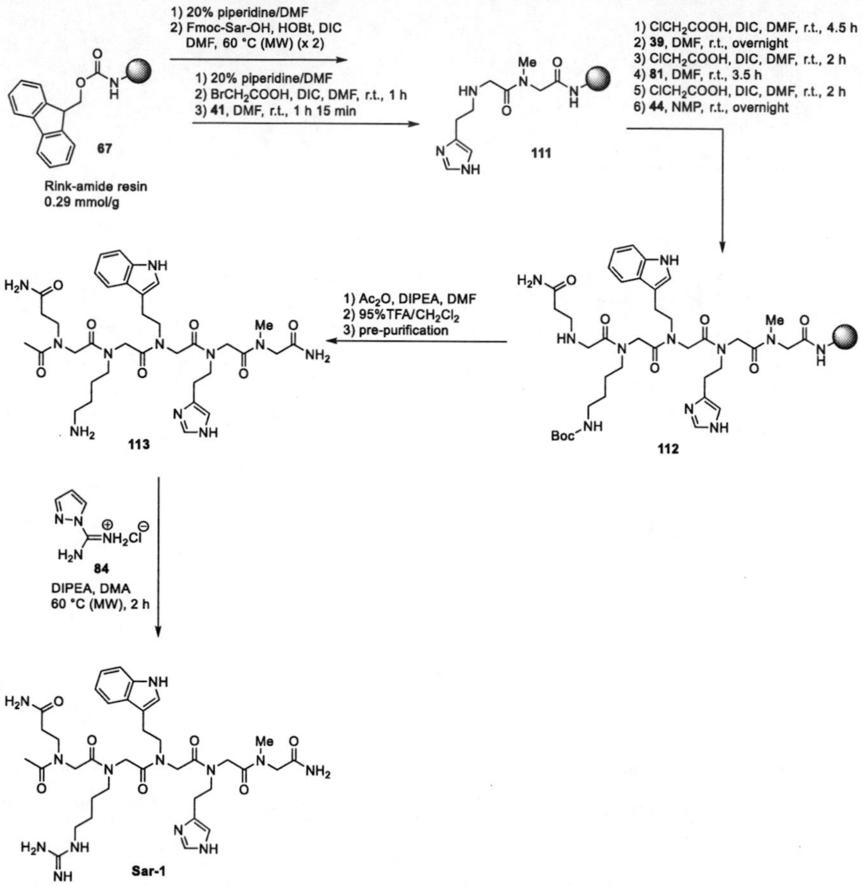

Scheme 25: Synthesis of **Sar-1** on low-loading Rink-amide resin. This route was used as a model for the synthesis of the other Sar-peptoids, modifying only the position of the sarcosine monomer.

The ability of the synthesized peptoids to inhibit Met-dependent migration of tumor cells was evaluated in a scratch-wound assay. The results are presented in Chart 4. Of the five modified peptoids tested in the sarcosine scan, only **Sar-1** retained the ability to block HGF-induced cell migration, indicating that the glutamate side-chain at the *C*-terminus is not essential for the inhibitory activity. The other Sar-peptoids did not show significant suppression of cell migration in comparison with the control, revealing the importance of each of the altered functionalities.

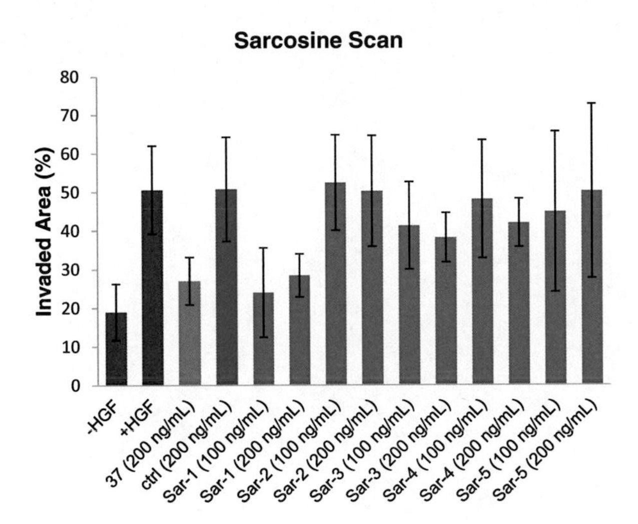

Chart 4: Summarized results of the scratch wound assays with L3.6pl cells for the sarcosine scan. The percentage of invaded area was measured after 24 h (n = 4). Purple: controls without peptides/peptoids. Green: CD44-v6 peptide (positive control) and control peptide. Blue: peptoids.

The conservation of the biological activity in **Sar-1**, as well the loss of activity for **Sar-2**, **Sar-3** and **Sar-4** (where the central residues have been mutated) are consistent with the results obtained in the peptide alanine scan. However, the inability of **Sar-5** to inhibit cell migration was unexpected, especially since peptoid **88**, also lacking the original asparagine side-chain at the *N*-terminus, was still able to block HGF-induced cell migration (see Chart 1). These intriguing findings led to a closer examination of the effect of the substituents at the *N*-terminus.

Effect of the substituents at the *N*-terminus

Aiming to understand the unexpected loss of activity of peptoid **Sar-5**, lacking the asparagine side-chain at the *N*-terminus, substitutions at this position were studied. A comparison of the *N*-terminus substituents of the different active peptides and peptoids with **Sar-5** is shown in Figure 18. It was hypothesized that all the blocking oligomers are able to act as hydrogen bond donors at this position (either through the side-chain or the backbone) whereas **Sar-5** is not. In order to confirm this hypothesis two analogs of **Sar-5** (**115** and **117**) were synthesized, which were also missing the asparagine side-chain but, unlike **Sar-5**, were still able to form hydrogen-bonds at the *N*-terminus (Scheme 26).

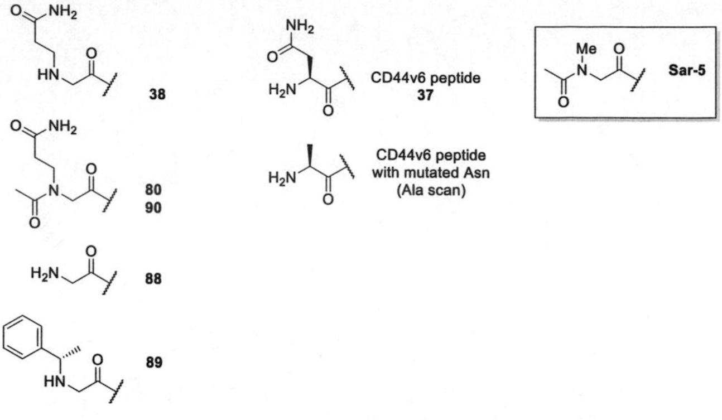

38
80
90
88
89
CD44v6 peptide
37
CD44v6 peptide
with mutated Asn
(Ala scan)
Sar-5

Figure 18: Comparison of the substituents at the *N*-terminus of the different active peptoids and peptides with **Sar-5**.

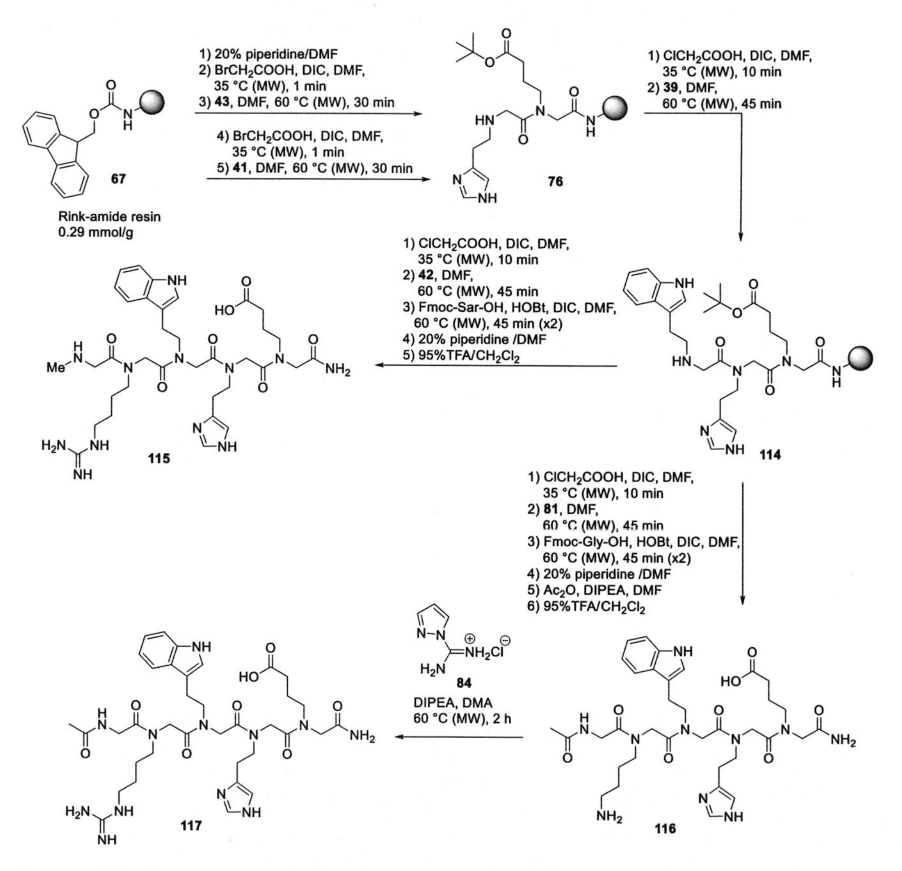

Scheme 26: Synthesis of peptoids **115** and **117** on low-loading Rink-amide resin.

The ability of peptoids **115** and **117** to inhibit HGF-induced cell-migration was evaluated in a scratch-wound assay (Chart 5). Unfortunately, none of the peptoids significantly blocked cell-migration in comparison with the control and the hypothesis of the hydrogen-bond donor at the *N*-terminus was discarded. Further investigations are necessary to understand the role of the substituents at this position.

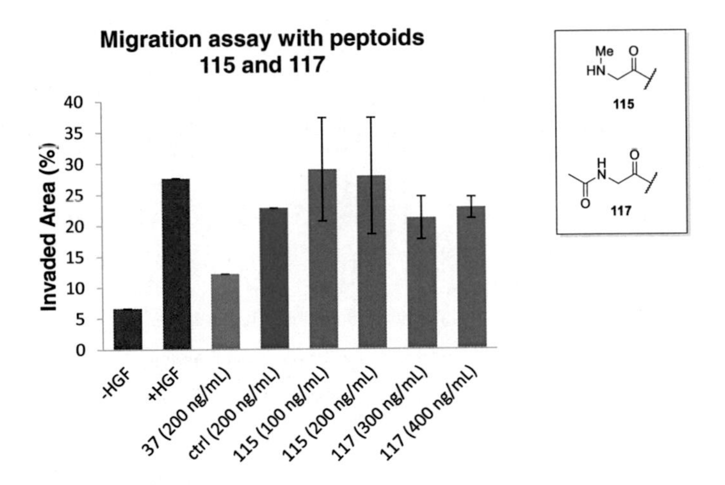

Chart 5: Summarized results of the scratch wound assays with L3.6pl cells for peptoids 115 and 117. The percentage of invaded area was measured after 24 h. Purple: controls without peptides/peptoids. Green: CD44-v6 peptide (**37**, positive control) and control peptide. Blue: peptoids.

SAR Summary for the human CD44-v6 peptoids

Comparing the structures of all the peptoids that were able to block HGF-induced Met activation in the described assays, it is possible to identify some essential features for activity.

118

R^1 = H, Me, $(CH_2)_2COOH$, $CH(CH_3)Ph$
R^2 = NH_2, $NHC(=NH)NH_2$
R^3 = $(CH_2)_2CONH_2$, $CH(CH_3)Ph$, H (if R^4 = H)
R^4 = H, Ac

Figure 19: General representation of the Met-inhibiting CD44-v6 linear peptoids.

- Substitution at position R^1 (Figure 19) is not critical for activity and many different residues are tolerated (see peptoids **38, Sar-1, 88** and **89**).

- The imidazolyl (N2im) and indolyl (N2in) side-chains seem to be essential for the inhibitory effect, as revealed by the lack of activity of peptoids **Sar-2** and **Sar-3**. Tolerance for different heterocycles and ring substituents will be investigated in the future.

- The guanidine functionality of the original peptide at R^2 (Figure 19) could be exchanged for an amine without loss of activity (see peptoid **88**), but replacement of the side-chain by a methyl group rendered an inactive compound (see peptoid **Sar-4**). A protonatable residue is probably needed at this position.

- The amide side-chain at the *N*-terminus (Figure 19, R^3) is not crucial for activity, since peptoids **88**, **89** and cyclic peptoids **91** and **92** were still able to inhibit scattering and migration of tumor cells. However not every substituent is tolerated (see peptoids **Sar-5**, **115** and **117**). The role of the substituent at the *N*-terminus needs to be further investigated.

- Conformational rigidity and orientation of the side-chains does not seem to be a critical parameter for the activity of CD44-v6 peptoids, since both linear and cyclic peptoids of different ring sizes, were able to block HGF-induced scattering and migration of tumor cells compared to the controls. However the order of the side-chains from *C*- to *N*-terminus in the linear peptoids does seem to be important as indicated by the loss of activity of retropeptoid **87**.

It is important to mention that these experiments are only qualitative and do not allow to quantify small variations in the efficiency of the different active peptoids. The general guidelines described above could be used to design new Met-inhibiting peptoids. However, the selection of the optimal candidate would need additional experiments. Moreover, pharmacokinetic properties, solubility and toxicity should also be taken into account.

4.2.7 **In vivo** *experiments*

Peptoid **38** which showed the strongest inhibition within the linear peptoids in the migration assay (Chapter 4.2.5, Chart 1), was selected for an *in vivo* experiment.

In this experiment nude mice, previously injected with L3.6pl human pancreatic carcinoma cells, were treated with peptoid **38** (treatment group) or control peptoid **107** (control group) three times a week during 21 days.

There was no difference in the size and weight of the primary tumor between the treatment and control groups (Figure 20, Charts A and B). However, there was a significant decrease in the amount of blood vessels in the primary tumor for mice treated with peptoid **38** (Figure 20, Chart C), indicating slower tumor growth.

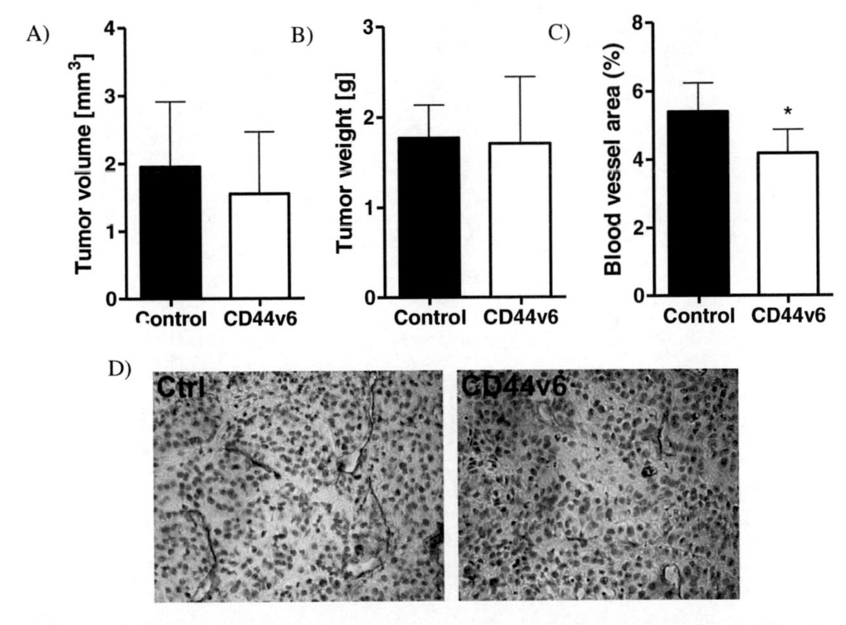

Figure 20: A) Comparison of the tumor volume between mice treated with CD44-v6 peptoid **38** and control peptoid **107**. B) Comparison of the tumor weight between mice treated with CD44-v6 peptoid **38** and control peptoid **107**. C) Comparison of the amount of blood vessels in the primary tumor between mice treated with CD44-v6 peptoid **38** and control peptoid **107**. A mouse with an area of blood vessels much higher than the rest was excluded from the control group. ($n = 4, 5$; $p < 0.05$) D) Pictures of CD31 staining of the primary tumor for the determination of blood vessels. Left: mice treated with control peptoid **107**. Right: mice treated with the CD44-v6 peptoid **38**.

Moreover, a significant reduction in the number of macroscopic liver metastasis was found in mice treated with peptoid **38** in comparison with the control (Figure 21). Considering that metastasis is the main cause for the high morbidity in cancer patients,[125] the ability of peptoid **38** to reduce the number of liver metastatic events *in vivo*, makes it a promising candidate for tumor therapy.

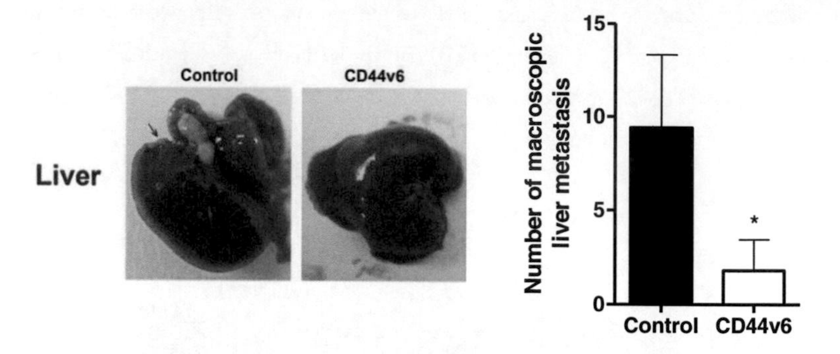

Figure 21: Comparison of the macroscopic liver metastases between the mice treated with peptoid **38** and control peptoid **107**. (n = 5; $p < 0.05$). The pictures on the left show two examples of livers from the control and treatment groups, where the metastases have been marked with a red arrow.

4.3 BAG-1 Peptoids

The BAG-1 peptide RVMLIGK (**119**) was the second short peptide oligomer chosen for the development of peptoid-based peptidomimetics. As mentioned in the introduction, this heptamer in the BAG-1 protein is essential for its binding to the cancer-related proteins GRP75 and GRP78.[109]

The BAG-1 peptide **119** and the peptoid analog **120** bearing the same side-chains and in the same order as the peptide, are depicted in Figure 22. As described for the CD44-v6 peptoids, the side-chains in the target structure are one methylene unit longer than the original peptide residues. This small change allows the use of biogenic amines as submonomers, which are often commercially available.

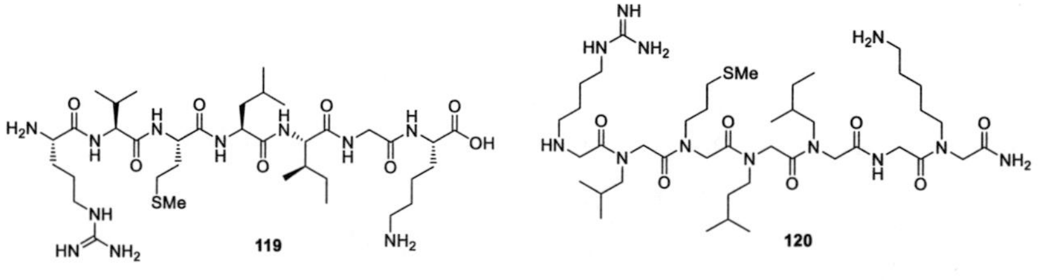

Figure 22: Chemical structures of the BAG-1 heptapeptide (**119**) and its peptoid analog **120**.

4.3.1 Submonomer synthesis

The amines needed for the synthesis of the target peptoid **120** *via* the submonomer method are represented in Figure 23.

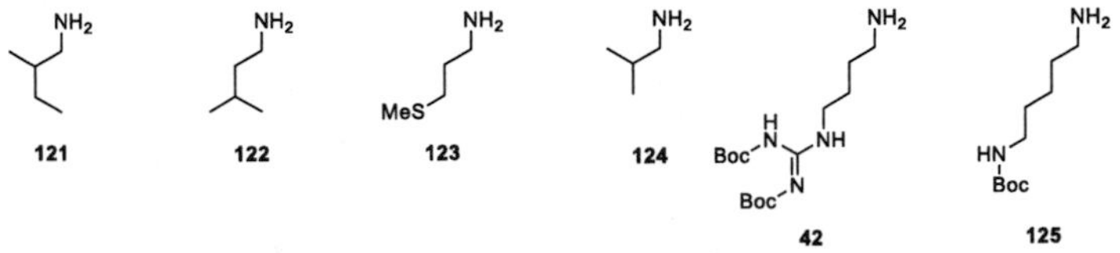

Figure 23: Amine submonomers needed for the synthesis of target peptoid **120**.

Amines **121-124** are commercially available. Submonomer **42** was already used in the assembly of CD44-v6 peptoids and its preparation has been described in Chapter 4.2.1. Mono-protected diamine **125** was synthesized in high yield as described in Scheme 27.

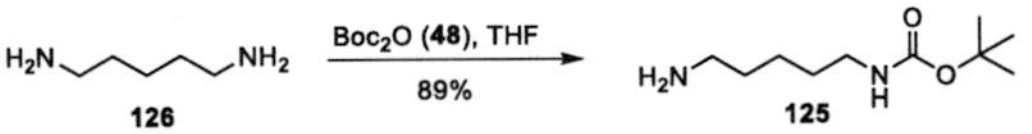

Scheme 27: Synthesis of Boc-protected submonomer **125**.

The amino acid attached to the *N*-terminus of **119** in the original BAG-1 protein is a cysteine. This side-chain was also introduced in some of the peptoid analogs. For its incorporation into peptoids the thiol functionality was protected with the acid-labile monomethoxytrityl (Mmt) group. Protection of cysteamine hydrochloride (**127**) was performed following a procedure described by Valliant and coworkers (Scheme 28).[126] Thus, amine **128**, needed for the introduction of the cysteine side-chain *via* the submonomer method, could be obtained in 75% yield.

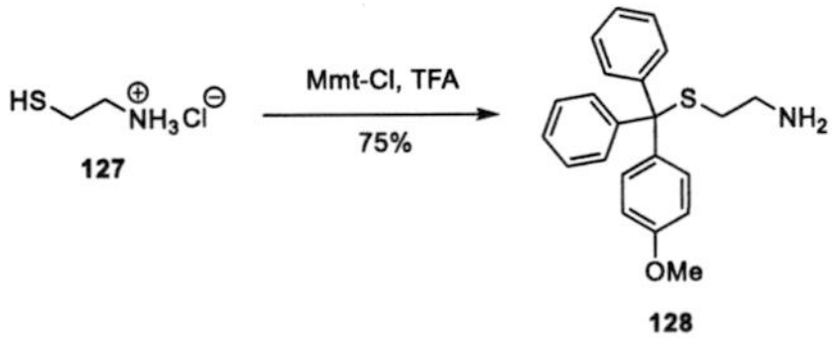

Scheme 28: Synthesis of submonomer **128**.

4.3.2 Synthesis of peptoid analogs of the BAG-1 peptide

With the submonomers in hand, six linear peptoids mimicking the BAG-1 heptamer **119** were synthesized (Figure 24).

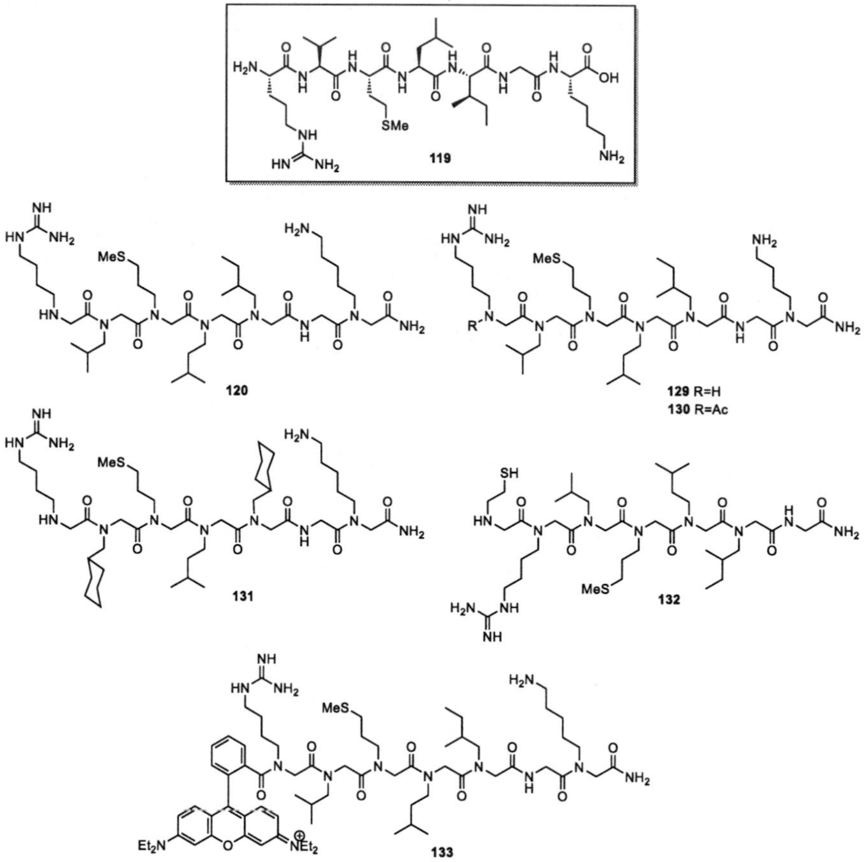

Figure 24: BAG-1 peptide (**119**) and synthesized linear peptoid analogs.

In addition to target peptoid **120**, a few peptoids with small side-chain modifications were also prepared. In peptoids **129** and **130** the lysine side-chain was shortened by one methylene unit. In addition, peptoid **130** was acetylated at the *N*-terminus. In peptoid **131** some branched alkyl side-chains were exchanged by cyclohexyl residues. Peptoid **132** lacks the lysine side-chain at the *C*-terminus and instead a new side-chain has been attached at the *N*-terminus, thus formally shifting the target heptamer one unit along the BAG-1 protein. Finally, rhodamine-labeled peptoid **133**, bearing the same side-chains as the original peptide (like peptoid **120**), was synthesized for localization experiments.

The BAG-1 peptoids in Figure 24 were initially synthesized on high loading Rink-amide resin (0.69 mmol/g). All residues, except glycine, were introduced following typical submonomer method protocols with bromoacetic acid as acylating reagent (see Chapter 2.2.1, Scheme 3). The HPLC profiles of the crude peptoids were considerably purer than those obtained for the CD44-v6 peptoids (Figure 25). The byproducts were identified as dimers generated by the cross-linking reaction already described in Chapter 4.2.2 (Scheme 29).

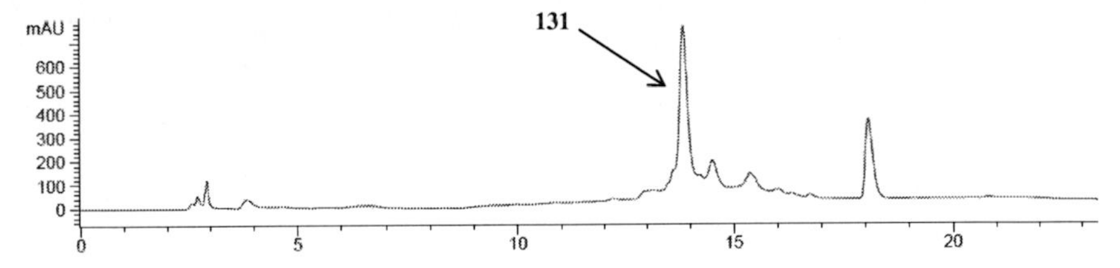

Figure 25: Chromatogram of crude peptoid **131** synthesized *via* typical submonomer protocols on high-loading Rink-amide resin (0.69 mmol/g). Chromatogram recorded at 218 nm. HPLC gradient: 5–95% acetonitrile in water + 0.1% TFA over 20 min. The x-axis indicates retention time in minutes.

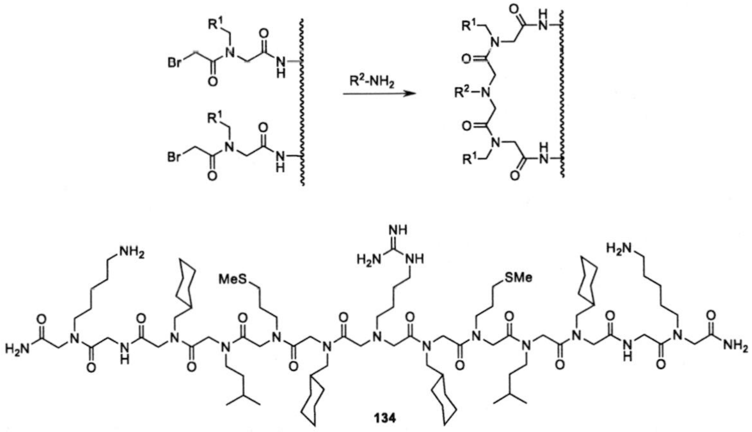

Scheme 29: Cross-linking side-reaction leading to dimer formation and example of dimerization byproduct **134** obtained in the synthesis of **131**.

The incorporation of guanidinated submonomer **42** at the end of the sequence, did not affect the purity as significantly as the introduction of this residue in the middle of the CD44-v6 peptoid oligomers discussed in Chapter 4.2.3, as demonstrated by the chromatogram of crude peptoid **131** (Figure 25). On the contrary, incorporation of unprotected diaminobutane (**51**) followed by guanidination on solid phase (as described for peptoid **80** in Chapter 4.2.3), here resulted in numerous side-products and difficult purifications, due to the higher resin loading (Scheme 30).

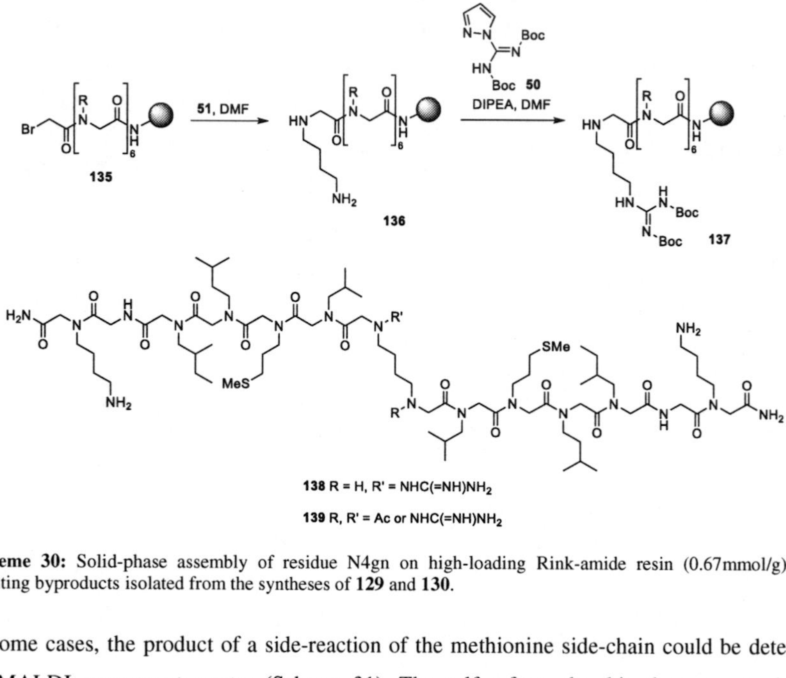

138 R = H, R' = NHC(=NH)NH$_2$

139 R, R' = Ac or NHC(=NH)NH$_2$

Scheme 30: Solid-phase assembly of residue N4gn on high-loading Rink-amide resin (0.67mmol/g) and resulting byproducts isolated from the syntheses of **129** and **130**.

In some cases, the product of a side-reaction of the methionine side-chain could be detected by MALDI mass-spectrometry (Scheme 31). The sulfur from the thioether can attack the bromide in an intramolecular nucleophilic substitution forming a seven-membered ring. This side-reaction had been previously observed in our group.[119] It could be prevented by using chloroacetic acid for the coupling after N3sm, which has a poorer leaving group than bromoacetic acid.

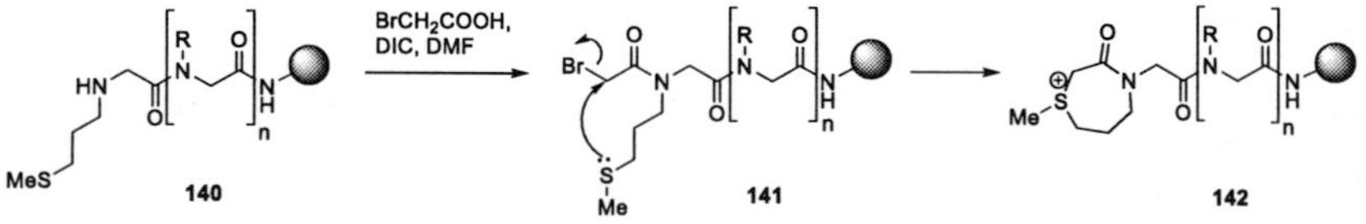

Scheme 31: Methylthio side-chains can undergo an intramolecular side-reaction.

In order to decrease the formation of the byproducts described above, rhodamine-labeled peptoid **133** was synthesized on low-loading Rink-amide resin (0.34 mmol/g), to reduce cross-linking, and chloroacetic acid was coupled after N3sm to prevent the intramolecular side-reaction of the thioether (Scheme 32). Thus, the desired peptoid was obtained in 10% yield over 16 steps.

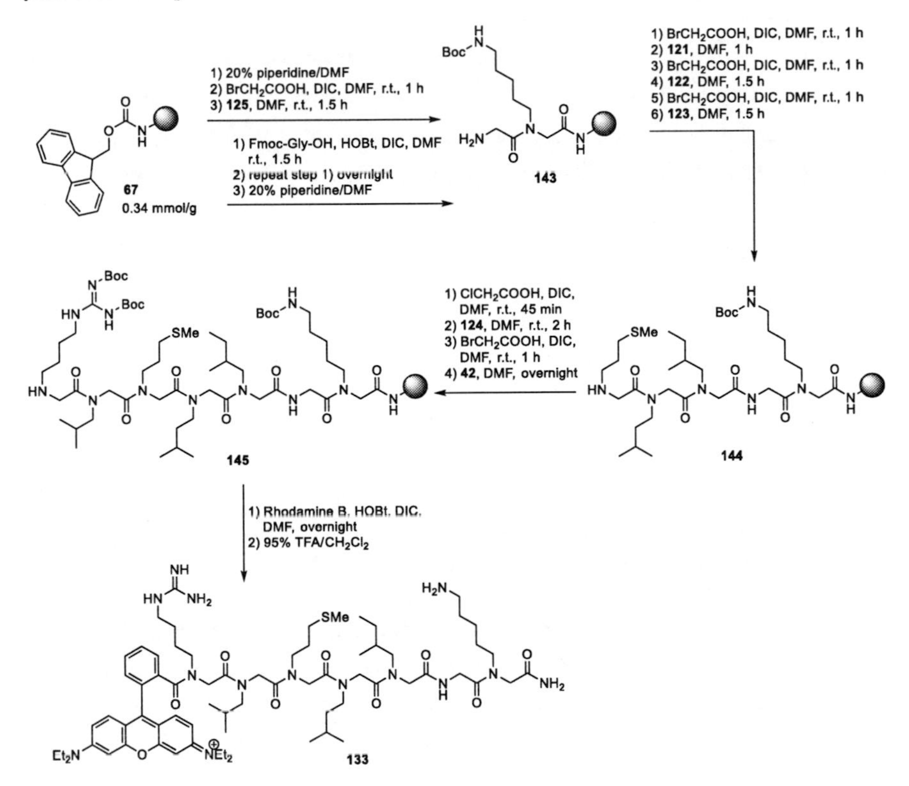

Scheme 32: Synthesis of rhodamine-labeled peptoid **133**.

4.3.3 Biological assays with BAG-1 peptoids

The biological assays described in the following chapters were performed by Dr. Antje Neeb in the research group of Prof. Cato.

Peptoids **120** and **129-132** were tested for inhibition of GRP78. Cato and coworkers had previously shown that stress-induced apoptosis (regulated by GRP78) was increased in cells expressing the BAG-1 peptide (68mer).[109] The peptide inhibits the refolding activity of GRP78 resulting in an enhanced susceptibility to apoptosis induced by the unfolded protein response (UPR).

The ability of the synthesized peptoids to increase stress-induced apoptosis in 22Rv.1 prostate cancer cells was evaluated. After incubation with the peptoids, the cells were treated with thapsigargin or glucose starved to induce ER stress and the expression of apoptotic markers (caspase 4 and PARP) was analyzed. Unfortunately no difference was observed between the cells incubated with the peptoids and the negative control (data not shown). This could be either due to the low level of peptoid taken up by the cells or due to accumulation of the peptoids in cellular compartments where the target receptors are not expressed. In order to investigate whether the peptoids bind to and are internalized by the prostate cancer cells, a fluorescent label was attached to the molecule, to allow detection and tracking of the peptoid in the cells.

4.3.4 Synthesis of fluorescein-labeled BAG-1 peptoids

For the fluorescence-based binding experiments (e.g. fluorescence activated cell sorting = FACS), the peptoids needed to be attached to a fluorescein dye. Therefore, a fluorescein-labeled peptoid with the same side-chains as the original BAG-1 peptide (**146**) and a control mutant peptoid (**154**), lacking some of the central side-chains, were synthesized.

Peptoid **146** was prepared from peptoid **145**, the synthesis of which has been described in Scheme 32. After elongation of the peptoid with an Ahx (aminohexanoic acid) spacer, fluorescein-isothiocyanate (FITC) was coupled to the *N*-terminus. Final cleavage from the resin afforded peptoid **146** (Scheme 33).

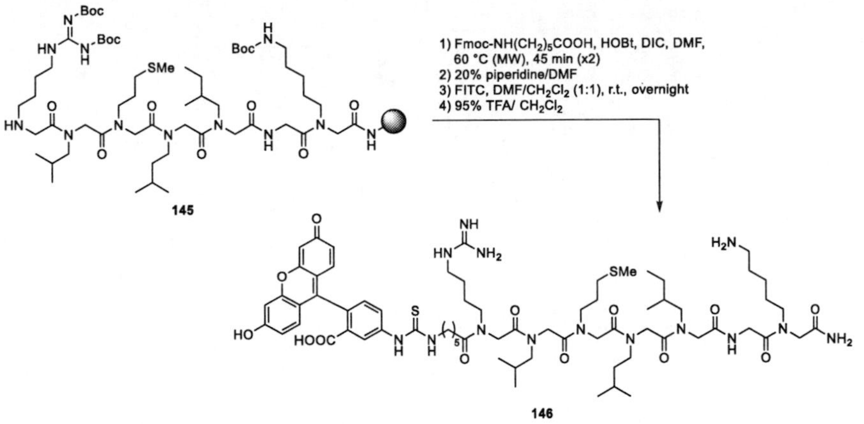
1) Fmoc-NH(CH2)5COOH, HOBt, DIC, DMF, 60 °C (MW), 45 min (x2)
2) 20% piperidine/DMF
3) FITC, DMF/CH2Cl2 (1:1), r.t., overnight
4) 95% TFA/ CH2Cl2
145
146

Scheme 33: Synthesis of fluorescein-labeled peptoid **146**. FITC = fluorescein isothiocyanate.

The spacer was introduced to avoid Edman degradation during acidic cleavage from the resin. This reaction, described in Scheme 34, was reported by Edman in 1950 as a method for peptide sequencing.[127]

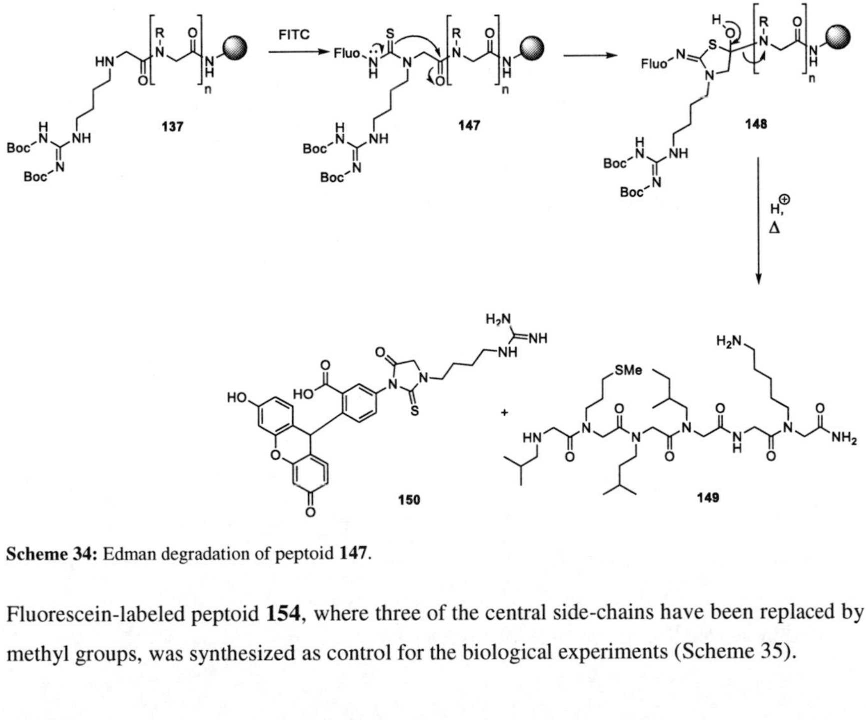
FITC
137
147
148
150
149

Scheme 34: Edman degradation of peptoid **147**.

Fluorescein-labeled peptoid **154**, where three of the central side-chains have been replaced by methyl groups, was synthesized as control for the biological experiments (Scheme 35).

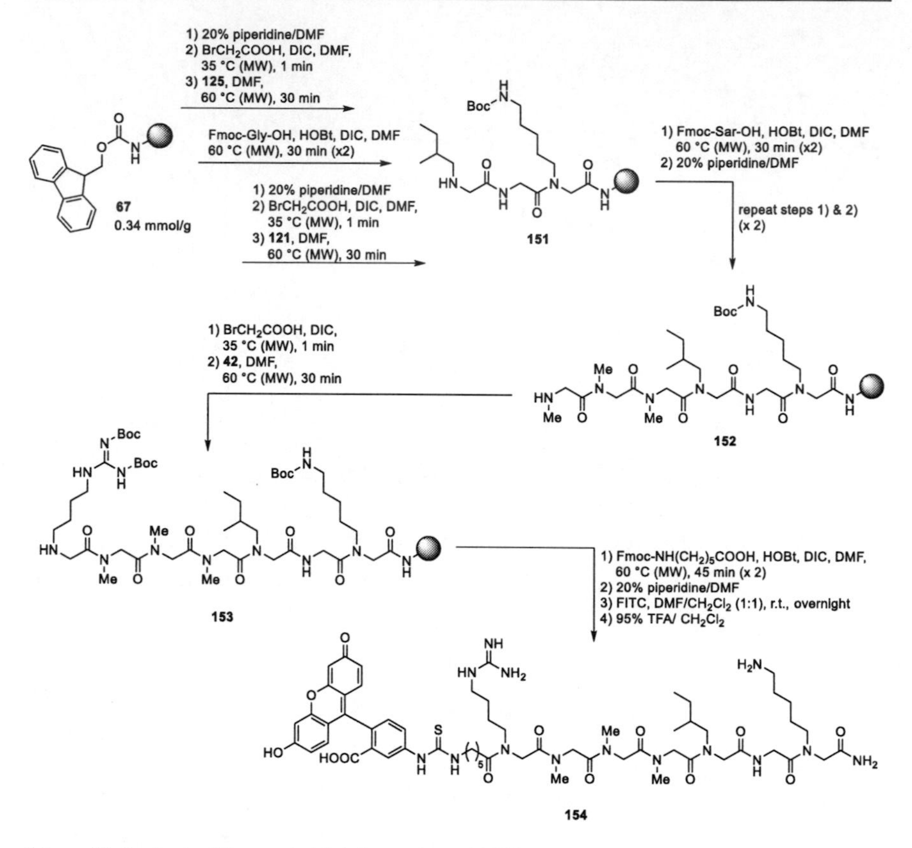

Scheme 35: Synthesis of fluorescein-labeled control peptoid **154**.

4.3.5 Cell binding assays

To study whether the peptoid binds to prostate cancer cells, 22Rv.11 cells that express high levels of GRP75 and GRP78, both intracellular as well as on the surface, were chosen as a model system. Utilizing these cells and fluorescein-labeled Bag-1 peptoid **146** a fluorescence activated cell sorting (FACS)-based binding assay was performed. 22Rv.1 prostate cancer cells were incubated with 0.1, 1.0, 10, 100 and 1000 nM peptoid **146** or mutant control **154** for 1 h at 4 °C. Then the cells were washed with PBS and analyzed via FACS. The result is shown in Figure 26. The purple histogram represents the untreated control cell population, the green lines show the fluorescence of the cells that have been incubated with either the fluorescein-labeled Bag-1 peptoid or the mutant control.

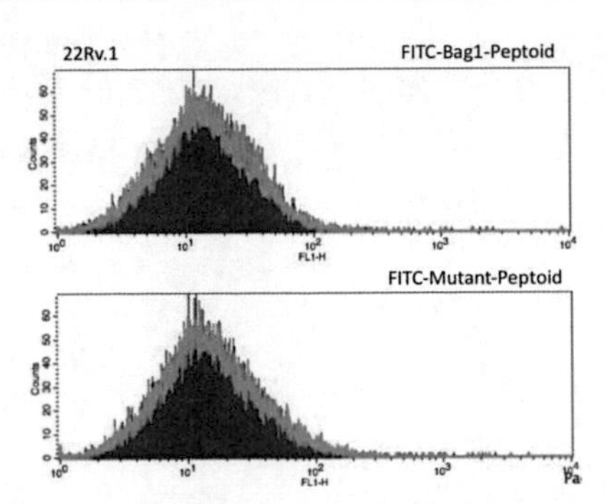

Figure 26: Results of FACS-based cell binding study. The untreated control cell population is represented in purple and the fluorescence of the cells incubated with the peptoids is shown in green. FITC-Bag1-Peptoid (top) = **146**. FITC-Mutant-Peptoid (bottom) = **154**.

Neither with the fluorescein-mutant peptoid nor with the fluorescein-labeled Bag-1 peptoid a shift in fluorescence intensity could be observed; indicating that either the peptide did not bind to the cells, the reduced time and temperature of the binding assay was insufficient to allow the peptoid to bind, or the fluorescence of the fluorescein-label was somehow quenched by the attachment of the peptoid. To evaluate these possibilities a rhodamine-labeled Bag-1 peptoid was utilized as a control and challenged together with the fluorescein-peptoids in a cell binding assay using attached living cells under standard culture conditions at 37 °C for 24 h. Briefly, tumor cells were seeded in full medium together with the rhodamine-labeled peptoid **133** for 24 h. After 24 h the cells were fixed and counterstained with an antibody against GRP75 and Draq5 to mark the position of the nuclei. Analysis of the cells using a confocal microscope confirmed the results of the FACS-experiments, since the fluorescein-labeled peptoid could not be detected inside or at the surface of the 22Rv.1 prostate cancer cells (Figure 27). The same experiment performed with the rhodamine-tagged peptoid, however, showed a different result (Figure 28). The peptoid (red) entered the cells mainly in vesicular structures and colocalized at least partially with GRP75 (green).

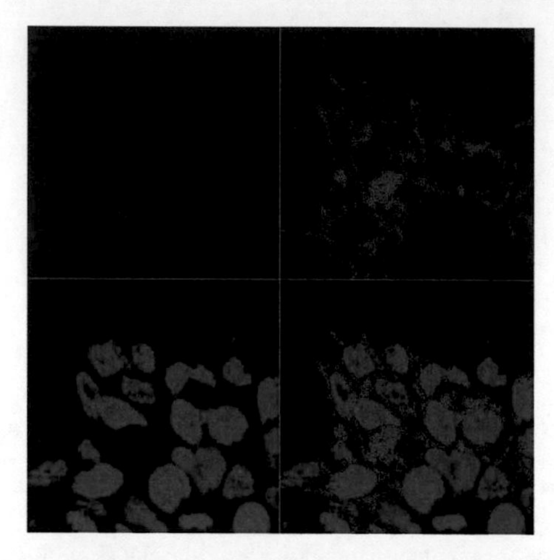

Figure 27: Confocal pictures of 22Rv.1 cells incubated with peptoid **146** (1 µM) for 20 h at 37 °C and fixed with paraformaldehyde. The cell nuclei were stained with Draq5 (blue) and GRP75 with a specific antibody (red). The fluorescein-labeled peptoid **146** should appear in green. The pictures show the green (top left), red (top right), and blue (bottom left) channels, as well as the merge of the three channels (bottom right).

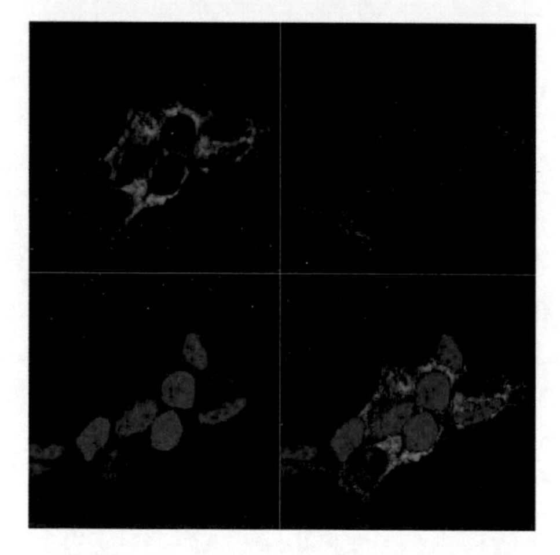

Figure 28: Confocal pictures of 22Rv.1 cells incubated with peptoid **133** (1 µM) for 20 h at 37 °C and fixed with paraformaldehyde. The cell nuclei were stained with Draq5 (blue) and GRP75 with a specific antibody (green). The rhodamine-labeled peptoid **133** (red) is located in vesicular structures. The pictures show the green (top left), red (top right), and blue (bottom left) channels, as well as the merge of the three channels (bottom right).

The fact that rhodamine-labeled peptoid **133** is able to penetrate inside the cells, while fluorescein-labeled peptoid **146** is not, is unexpected. Since rhodamine B by itself is not internalized, the explanation must be sought in the combined peptoid-rhodamine entity. It is possible that the amphiphilic structure generated by attaching the relative hydrophobic peptoid to the positively charged rhodamine, favors cell attachment and/or internalization. Further experiments would be necessary to understand this phenomenon.

4.4 Peptoid Oligomers with Chiral Amide Side-chains

4.4.1 Synthesis of peptoid oligomers with chiral amide side-chains

While investigating the incorporation in peptoid sequences of amide side-chains mimicking asparagine and glutamine (see Chapter 4.2.2), the scope was extended to chiral amide submonomers **155** and **156** (Figure 29). The interest for α-chiral side-chains lies in their ability to form stable secondary structures.[40] Residues **155** and **156** are so far unknown in peptoid chemistry, but substituted amide derivatives of **155** have been used in the synthesis of peptoid multihelical structures.[102]

H_2N O NH_2 Me
155

H_2N O NH_2 Ph
156

Figure 29: New chiral-amide peptoid submonomers.

Submonomer **155** could be synthesized from commercially available Cbz-protected succinimidyl activated L-alanine (**157**). Treatment of this compound with ammonia followed by deprotection with H_2 on Pd/C afforded the desired submonomer as a free amine in very good yields (Scheme 36). (*R*)-Phenylglycineamide (**156**) is commercially available.

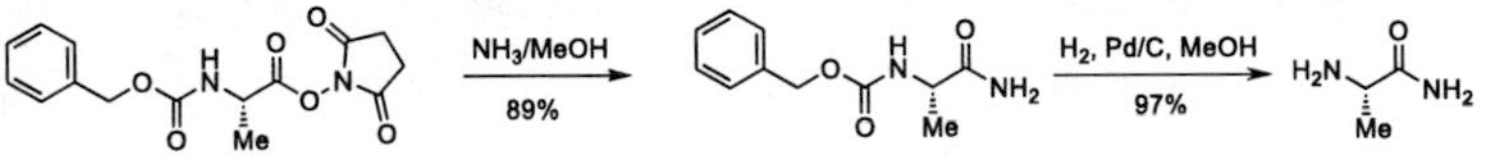

157 —(NH_3/MeOH, 89%)→ **158** —(H_2, Pd/C, MeOH, 97%)→ **155**

Scheme 36: Synthesis of submonomer **155**.[121]

First, the incorporation of the new side-chains in peptoid oligomers was studied in test sequences, as described for compounds **44** and **45** (Chapter 4.2.2). Both submonomers (**155** and **156**) were successfully coupled to the test peptoid **68** and the obtained oligomers could undergo a new coupling step to afford peptoids **159** and **160** (Scheme 37). However, this subsequent acylation was not complete and 26% of the trimer, lacking the last monomer was isolated in the synthesis of **159**. Moreover the final secondary amine in peptoids **159** and **160** reacted within hours with the amide side-chain to generate a diketopiperazine (DKP).

Therefore peptoids **159** and **160** could only be obtained as inseparable mixtures with their DKP-byproducts (**159-DKP** and **160-DKP**). DKP-Formation has been previously reported as side-reaction in dipeptoids.[128]

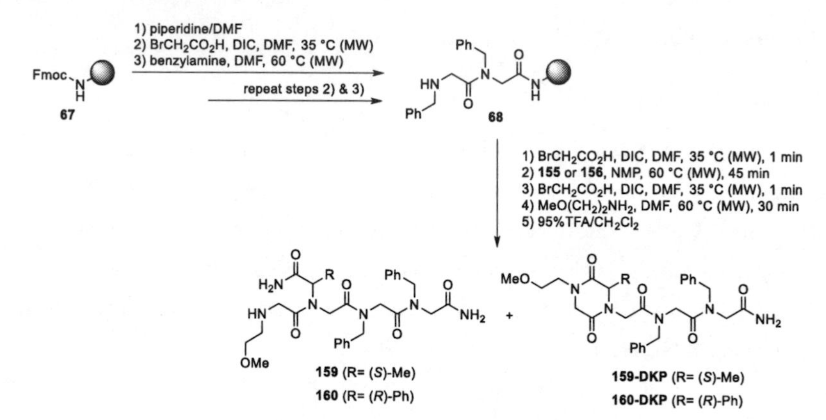

Scheme 37: Synthesis of test peptoid sequences **159** and **160**. Yield (**159**) = 22% inseparable mixture of **159** and **159-DKP**, 26% **159-DKP** and 26% trimer H-Nsmd-$(N1ph)_2$-NH_2. Yield (**160**): 41% inseparable mixture of **160** and **160-DKP**.

The combination of DKP-formation and incomplete acylation, due to the steric hindrance of submonomers **155** and **156**, led to difficult purifications and lower yields compared to the incorporation of unbranched amide-submonomers **44** and **45** (Chapter 4.2.2).

The limitation of DKP-formation for the use of α-chiral unsubstituted amides as peptoid side-chains was next investigated. The first approach was to immediately acetylate the peptoid *N*-terminus after substitution with the last submonomer (Table 4, route A). However under the capping conditions (Ac_2O/DIPEA/DMF) DKP-formation was faster than acetylation and an approximately 2:1 **DKP-160**/product (**162**) ratio in the crude mixture was detected by HPLC (Table 4, entry 1). Moreover, acylation after the amide residue was incomplete and the starting material **161** represented 27% of the HPLC profile. Elongation of the peptoid with a fifth coupling (Table 4, route B), afforded a similar **DKP-160**/product (**163**) ratio (Table 4, entry 2) indicating that DKP-formation is fast and probably takes place during the substitution step with methoxyethylamine. In contrast with route A, here the double acylation after the amide submonomer allowed complete conversion of starting material **161**. If the submonomer following the α-chiral amide side-chain is a secondary amine, DKP-formation is not possible. As proof-of-principle compound **164** was synthesized (Table 4, route C) and the desired product was successfully obtained in 67% HPLC purity in the crude mixture (Table 4, entry 3).

Table 4: Studies for DKP-formation in peptoids containing α-chiral side-chains with unsubstituted amides.

A: 1) $BrCH_2COOH$, DIC, DMF, 35 °C (MW), 1 min; 2) $MeO(CH_2)_2NH_2$, DMF, 60 °C (MW), 30 min; 3) Ac_2O, DIPEA, DMF

B: 1) $BrCH_2COOH$, DIC, DMF, 35 °C (MW), 1 min (x2); 2) $MeO(CH_2)_2NH_2$, DMF, 60 °C (MW), 30 min; 3) repeat steps 1) & 2)

C: 1) $BrCH_2COOH$, DIC, DMF, 35 °C (MW), 1 min (x2); 2) morpholine, DMF, 60 °C (MW), 30 min (x2)

161 → 162 (A), 163 (B), 164 (C)

Entry	Acylation	Route	Product	Product (%) in crude[a]	160-DKP (%) in crude[a]
1	1×	A	**162**[b]	12[b]	27
2	2×	B	**163**	28	53
3	2×	C	**164**	67	-

[a]Determined by HPLC, detection at 218 nm. [b]27% of starting material **161** was also detected by HPLC.

A possible improvement for the incorporation of α-chiral primary amides into peptoids, which could decrease the amount of DKP-byproduct, would be the coupling of the subsequent residue as an Fmoc-protected monomer.

To further investigate the coupling potential of the chiral amide submonomers (**155** and **156**), different length homo-oligomers were synthesized (Scheme 38). Here the acylation step after each submonomer was repeated twice (double coupling) to ensure completion of the reaction. DKP-formation in these structures was much slower than in the test sequences in Scheme 37, probably due to steric effects. Here it was possible to isolate the desired compounds, and the DKP-byproducts were only isolated in minor yields.

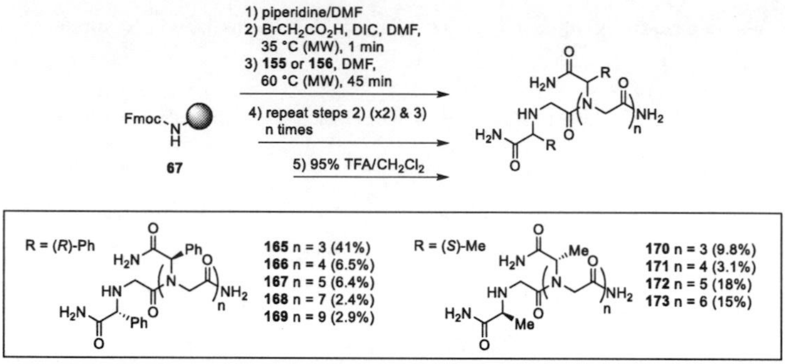

Scheme 38: Synthesis of peptoid homo-oligomers composed of α-chiral amide side-chains and yields of the isolated compounds after HPLC purification.

In general, the yields decrease with increasing chain length, due to the higher number of steps and the difficulty of the purifications. The anomalous low yields obtained for the short α-methyl substituted peptoids are due to the high polarity of these compounds, which required an initial adjustment of the HPLC purification conditions. Nevertheless these results prove that consecutive incorporation of the new chiral amide submonomers **155** and **156** in peptoid sequences is possible.

4.4.2 Structural investigations of peptoid oligomers with chiral amide side-chains

Peptoids composed of α-chiral side-chains are known to form stable helices.[40] Therefore the ability of the synthesized peptoid oligomers to adopt defined secondary structures was investigated.

A common technique to identify secondary structures in peptoids with chiral side-chains is circular dichroism (CD) spectroscopy. It is a simpler and faster tool compared to the complicated and time consuming characterization of peptoids by NMR spectroscopy due to the *cis/trans* amide equilibrium. However, it is not quantitative and conclusions can only be drawn by comparison with the spectra of model structures.

CD is defined as "the unequal absorption of left-handed and right-handed circularly polarized light".[129] A beam of light is associated to an electric and a magnetic field that oscillate perpendicular to each other. Light can be polarized so that the electric field oscillates sinusoidally in a plane. Viewed from the front the tip of the electric field vector moves up and down on the y-axis. This vector can be defined as the sum of two vectors of the same magnitude that describe circles in clockwise (R) and counterclockwise (L) directions (Figure 30, A).[130] If the light interacts with an asymmetric molecule that absorbs right- and left-handed polarized light differently, the addition of the two vectors (R'+L') results in an ellipse (Figure 30, B).[130] This is known as elliptically polarized light and the angle α is the optical rotation of the molecule. CD can be given as Δε, which is the difference between the extinction coefficients ε_R and ε_L, or ellipticity degrees, which is the angle (θ) whose tangent is the ratio of the minor to the major axis of the ellipse (Figure 30, C).

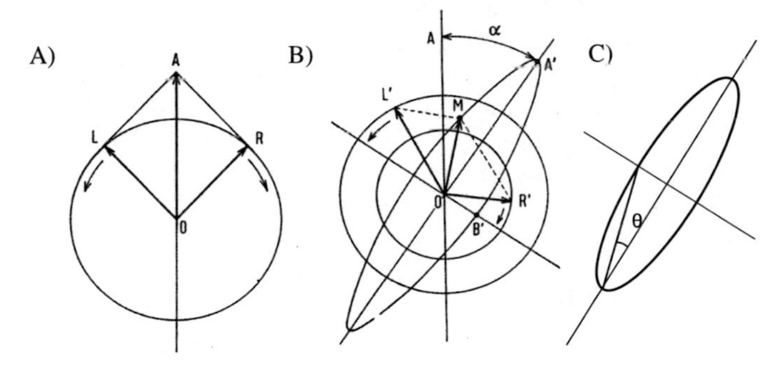

Figure 30: Circular dichroism principles and definitions. A) The electric vector of linearly polarized light can be described as a sum of two vectors that rotate clockwise (R) and counterclockwise (L).[130] B) Action of an optically active substance (absorbing right circularly polarized light to a greater extent) on plane polarized light.[130] C) Definition of ellipticity (θ).

In a CD spectrum the ellipticity is plotted against the excitation wavelength. In order to be able to measure a CD spectrum, the molecule needs to have chromophores that absorb light at the wavelength range used for the measurement (amide bonds for peptides and peptoids) and these chromophores need to be adjacent to a chiral center, for the molecule to be optically active. Interactions between chromophores in physical proximity cause changes in their optical transitions. This results in characteristic CD spectra for different secondary structures.[129]

The CD spectra of the methyl substituted Nsmd homo-oligomers could not be recorded due to solubility problems. The CD spectra of the phenyl substituted Nrpd peptoid homo-oligomers **166-169** are shown in Charts 6 A and B. On the y axis CD is given as mean residue ellipticity (MRE, Equation 1).

$$MRE\ (deg \cdot cm^2 \cdot dmol^{-1}) = \frac{\theta}{10 \cdot C_r \cdot d}$$

Equation 1: Mean residue ellipticity (MRE). θ = observed ellipticity (mdeg); C_r = molar concentration per peptide bond (mol/L), d = length of the cuvette (cm).

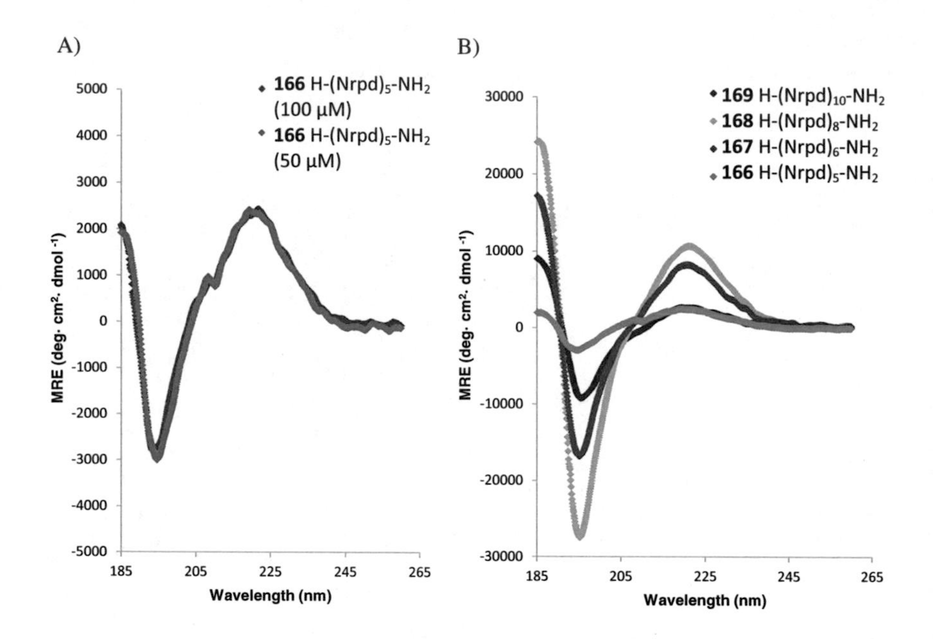

Chart 6: CD spectra of the synthesized α-chiral peptoids in acetonitrile. MRE = mean residue ellipticity. A) CD spectra of peptoid **166** at different concentrations. B) CD spectra of peptoids **166-169** as 50 µM solutions in acetonitrile.

In Chart 6, A the spectra of the Nrpd pentamer at two different concentrations are superposed to show that the shape and intensity of the CD signal is independent of the concentration. In Chart 6, B the CD signal of the different length oligomers can be compared. The spectra show a maximum around 220 nm, a shoulder around 205 nm and a minimum around 193 nm. The increasing intensity of the signals from pentamer to hexamer to octamer might suggest an increasing degree of ordering. The decamer, however, is left out of this tendency. Aggregation or solubility effects could be responsible for the weaker signal of $(Nrpd)_{10}$. Experiments at different concentrations with long oligomers (decamer and beyond) would be needed for a better understanding of this phenomenon.

The spectra in Chart 6 show some similarities with the CD spectra from peptoids rich in (*R*)-1-phenylethyl side-chains (e.g. **174**, Figure 31).[40] These spectra are the mirror image of α-helix type CD signals. The latter, are observed in peptoids rich in (*S*)-1-phenylethyl side-chains (e.g. **175**, Figure 31), and present two minima of negative ellipticity at 218 and 202 nm and a maximum of positive ellipticity at 190 nm.[40,131] In the synthesized Nrpd oligomers, however, the reported band around 202 nm has been reduced to a shoulder.

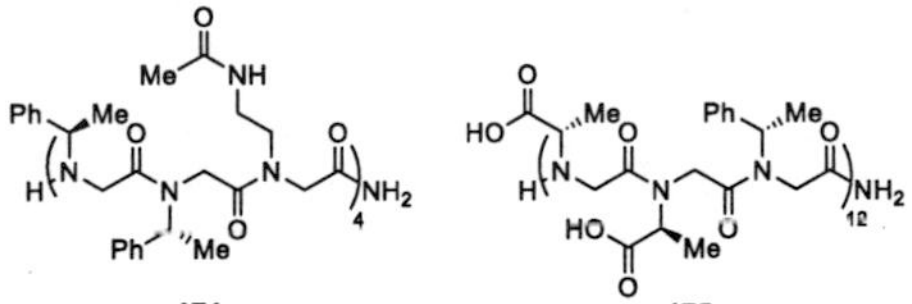

Figure 31: Examples of peptoid oligomers with α-chiral phenyl side-chains, which are able to adopt helical structures in solution.[40,131]

In conclusion, the similarities with previously reported CD-spectra of peptoid helices with α-chiral phenyl side-chains, especially the band around 220 nm, seem to indicate some degree of helicity in the synthesized peptoids. Nevertheless, the differences, both in the CD spectra and in the chemical structures, do not allow drawing further conclusions.

5. Summary and Outlook

Peptides are involved in many essential biochemical processes in living organisms and are therefore a very appealing class of compounds for biological and medical applications. However, their fast degradation by proteases *in vivo* limits their use as therapeutic agents. Thus, there is a great interest in the development of peptidomimetics able to carry out peptide functions with improved stability and bioavailability. Among these peptidomimetics, peptoids have attracted special attention thanks to their easily accessible and derivatizable synthesis on solid supports.

The aim of this work was to develop peptoid-based peptidomimetics of two short amino acid sequences with antitumor activity.

5.1 CD44-v6 Peptoids

The pentapeptide NRWHE (**37**) inhibits ligand dependent Met activation and subsequent migration of tumor cells by competing with the coreceptor function of CD44-v6.

A small library of peptoids mimicking the antitumor peptide **37** was synthesized and purified (selected examples are shown in Figure 32). Solid-phase synthesis of highly functionalized peptoids often leads to low yields and difficult purifications. However the optimization of the reaction conditions allowed for the synthesis of a variety of analogs. It is crucial that the submonomers be introduced as free bases. Moreover, the use of chloroacetic acid as acylating agent and low-loading resins significantly improved the purity of the synthesized peptoids. Derivatives with reduced flexibility could be obtained by incorporating α-chiral submonomers, peptide-peptoid hybrids or cyclization (e.g. **90**, **91**, Figure 32).

Many of the synthesized peptoids were able to inhibit HGF induced Met activation and block scattering and migration of tumor cells like the original peptide, showing their potential as antitumor agents. The strongest inhibitory effect was observed for peptoid **38**, with the same side-chains and in the same order as the peptide (Figure 32). Moreover, in an *in vivo* trial this peptoid was able to reduce the number of liver metastatic events and exhibited a decrease in angiogenesis, compared to a control peptoid. These results make peptoid **38** a promising candidate for cancer therapy.

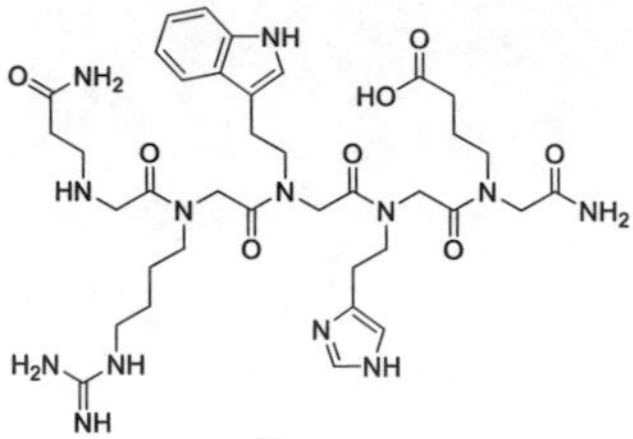

38

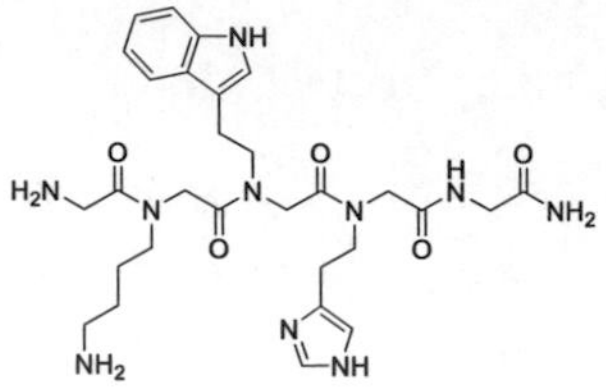

88

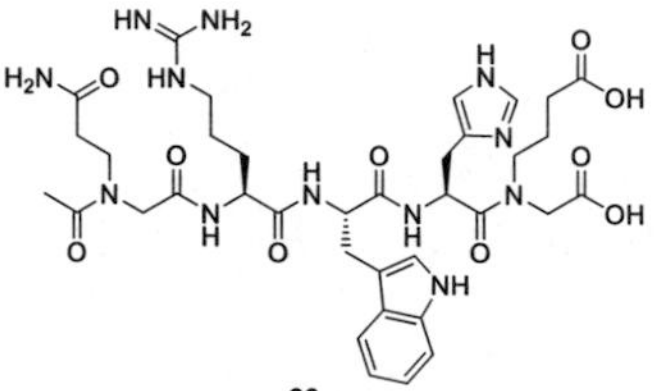

90

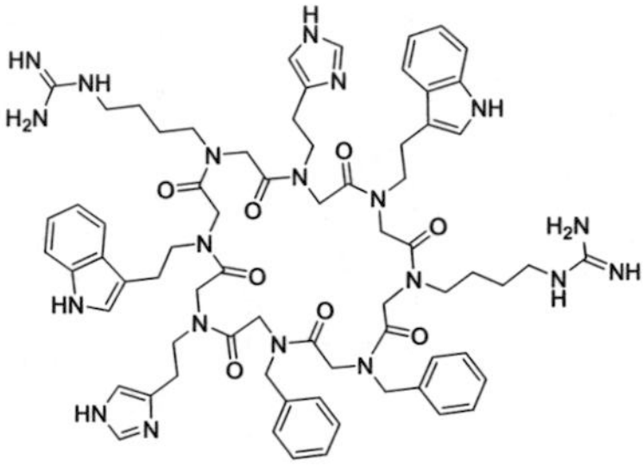

91

Figure 32: Selected examples of peptoid analogs of the human CD44-v6 peptide **37**.

Finally, a sarcosine scan, analogous to the alanine scan in peptides, was performed in order to determine which side-chains are responsible for the inhibitory activity. Altogether, comparing the structures and activities of the tested peptoids it was possible to establish some structure-activity relationships, which have been summarized in Figure 33.

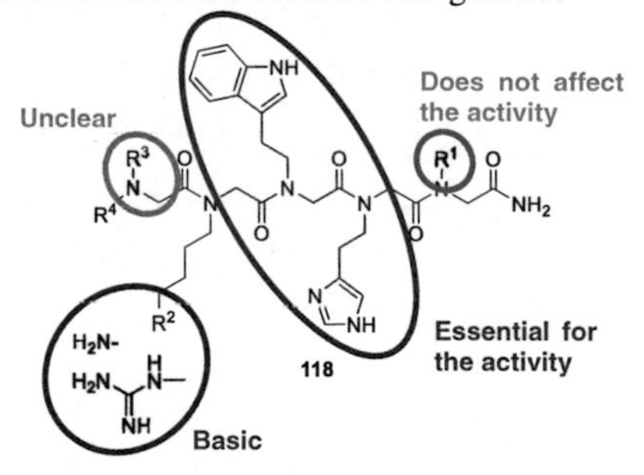

Figure 33: Structure-activity relationships in linear CD44-v6 peptoids.

As reported for the original peptide, the central peptoid side-chains (N2im, N2in and N4gn) seem to be essential for the inhibitory effect. However, the guanidine functionality could be exchanged for another basic residue without loss of activity. The role of the residue at the *N*-terminus is not clear yet and needs to be further investigated. While the amide functionality does not seem to be necessary, not every substituent is tolerated at this position.

Linear (e.g **38, 88-90**) as well as cyclic peptoids of different ring sizes (**191** and **192**) were able to block HGF-induced scattering and migration of tumor cells, suggesting that conformational rigidity and orientation of the side-chains is not a critical parameter. However, the order of the side-chains from *C*- to *N*-terminus in the linear peptoids does seem to be important, as indicated by the loss of activity of retropeptoid **87**.

In the future, the optimized reaction conditions could be applied to the synthesis of new analogs as probes for further SAR studies. The resulting information could then be used not only to design new Met-inhibiting peptoids, but also to identify the binding site *via* modeling studies, which would, at the same time, further contribute to the design of new inhibitors as potential antitumor agents.

The relative tolerance for different substituents at the *C*- and *N*-terminus could be used to tune solubility and pharmacokinetic properties. In cyclic peptoids, the residues pointing to one of the faces of the macrocycle can also be modified for that purpose. Furthermore it would be of great interest to synthesize fluorescently labeled peptoid analogs in order to follow their fate in the body and visualize their accumulation in tumor cells.

5.2 BAG-1 Peptoids

A peptide derived from the sequence of the cancer associated protein BAG-1 induces apoptosis and reduces tumor growth when overexpressed in prostate cancer cells, by binding the heat shock proteins GRP75 and GRP78. In order to use this effect for the development of new therapeutic inhibitors for the treatment of prostate cancer, several peptoid analogs of the heptapeptide RVMLIGK (**119**), essential for the binding, were synthesized (e.g. **120,** Figure 34).

The synthesized analogs were tested for inhibition of GRP78 by evaluating their ability to increase stress-induced apoptosis of 22Rv-1 prostate tumor cells. Unfortunately, the peptoids did not show any effect on apoptosis compared to the control. Therefore, it was postulated that the peptoids were either not able to enter the cell or ended up in non-GRP78-expressing compartments.

In order to address this question, rhodamine and fluorescein-labeled peptoid analogs of the BAG-1 peptide (**133** and **146**) were synthesized and incubated with 22Rv.1 prostate cancer cells *in vitro*. Surprisingly the fluorescein-labeled peptoid could not be detected, neither inside the cell nor at the surface. However, the rhodamine-labeled peptoid **133** was internalized by the cells in vesicular compartments and showed partial colocalization with GRP75. The reason for the contradictory results of the two tags remains unclear.

In the future, co-localization of the rhodamine-labeled peptoid **133** with GRP78 needs to be investigated. Binding studies (e.g. fluorescence polarization assays) should be performed to assess the ability of the peptoid to bind GRP78 and GRP75. Finally, elongation of the BAG-1 peptoid with cell-penetrating or mitochondrial directing sequences developed in our group,[16,132] might allow selective targeting of one receptor with respect to the other.

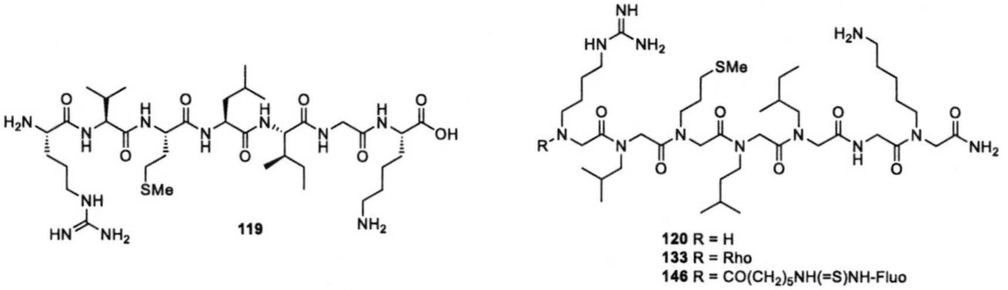

120 R = H
133 R = Rho
146 R = $CO(CH_2)_5NH(=S)NH$-Fluo

Figure 34: Chemical structure of the BAG-1 peptide **119** and selected peptoid analogs **120**, **133** and **146**.

5.3 Peptoid Oligomers with Chiral Amide Side-chains

Peptoids containing amide side-chains mimicking glutamine and asparagine are rare in literature and often lack detailed synthetic procedures. In this work a method for the synthesis of amide-containing submonomers as free bases and their incorporation into peptoid oligomers has been described. α-Chiral amide side-chains could also be successfully introduced in peptoid sequences. Furthermore, different-length homooligomers of chiral amide residues were synthesized demonstrating that consecutive incorporation of these residues is possible (Figure 35).

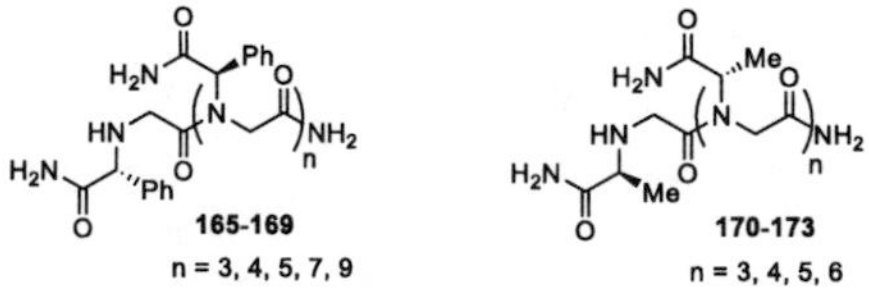

165-169 n = 3, 4, 5, 7, 9

170-173 n = 3, 4, 5, 6

Figure 35: Synthesized peptoid homooligomers composed of α-chiral amide residues.

The circular dichroism spectra of the synthesized phenyl-substituted Nrpd oligomers showed a certain degree of similarity to those of other peptoid sequences rich in α-chiral phenyl residues, which have been reported to adopt helical structures.

The method developed in this work allows the exploration of new amide-containing peptoids. The combination of the α-chiral amide side-chains with other residues improving the solubility and limiting the formation of side-products, may give rise to interesting secondary structures.

6. Experimental part

6.1 Materials and Methods

General

The starting materials were purchased from commercial sources (*Sigma Aldrich, Acros, ABCR, Merck*) and used without further purification. Analytical-grade solvents were also employed as purchased. Peptide-synthesis grade dimethylformamide (DMF) from *ABCR* was used for peptoid synthesis. DMF for peptoid cyclization was dried over sodium sulfate. Toluene and tetrahydrofuran were refluxed over sodium under argon atmosphere and distilled over a packed column. Dichloromethane was refluxed over calcium hydride under argon atmosphere and distilled over a packed column.

Moisture and air sensitive reactions were carried out under argon atmosphere according to the common Schlenck Technique.[133] Liquids were transferred *via* plastic syringes and V2A-steel needles. Powdered solids were added against an argon flow.

Reactions at 0 °C were cooled with an ice/water bath.

Solid phase syntheses were performed in 3, 6 and 8 mL glass or plastic fritted-syringes, closed with a plastic cap at the bottom. Rink Amide aminomethyl polystyrene resin (from *Merck-Novabiochem*) and 2-Chlorotrityl chloride polystyrene resin (from *Agilent* and *Carbolution*) were used as solid supports. Resin loading was given by the supplier and varied with every batch. It is specified for each reaction procedure.

Reaction control

Liquid-phase reactions were monitored by thin layer chromatography (TLC) on ready-to-use plates from *Merck* (silica gel 60 on aluminum sheet, F_{254} fluorescent indicator, 0.25 mm layer thickness). A UV lamp 204 AC from *Hanau Quarzlampen GmbH* with a wavelength of $\lambda = 254$ nm was used for visualization. Staining of amine compounds was accomplished by treatment with ninhydrin solution (0.2% in ethanol) followed by heating (aprox. 250 °C, heat gun). Seebach reagent (2.5% phosphomolybdic acid, 1.0% cerium (IV) sulfate tetrahydrate, 6.0% conc. sulfuric acid and 90.5% water), was also employed as developing agent.

For monitoring reactions on solid support, a test cleavage was performed on a few beads and the presence of the desired compounds was confirmed *via* MALDI-TOF mass spectroscopy. For the test cleavage the separated beads were treated with 20 μL trifluoroacetic acid (TFA), for Rink-amide resin, or hexafluoroisopropanol (HFIP), for chlorotrityl chloride resin, and shaken for 15 min at room temperature. Then 5–10 μL of the solution were taken, the solvent was evaporated and the residue was dissolved in the MALDI matrix.

For short peptoid sequences (under 500 g/mol) the chloranil test[134] was used as reaction control. Here a few beads of resin were separated and treated with 2–5 drops of 2% acetaldehyde solution in DMF and 2–5 drops of 2% chloranil solution in DMF. Dark blue to green beads indicate presence of secondary amines.

Column chromatography

Purification of the crude products by column chromatography was performed as described by Still.[135] Silica gel 60 (0.04–0.063 mm) from *Merck* was used as the stationary phase. The solvents for elution were purchased in analytical-grade purity and the mixtures are reported in volume ratios (v/v).

6.2 Analytical Methods and Instrumentation

Nuclear magnetic resonance spectroscopy (NMR)

NMR spectra were recorded at 25 °C on a *Bruker* AC 250 [250 MHz (^{1}H)], *Bruker* Avance 300 [300 MHz (^{1}H) and 75 MHz (^{13}C)] and a Bruker AM 400 [100 MHz (^{13}C)] spectrometer. Deuterated solvents were purchased from *Eurisotop*. Chemical shifts (δ) are expressed in parts per million (ppm) downfield from tetramethylsilane and referenced to the residual protons of the solvent: 7.26 ppm (^{1}H) or 77.16 ppm (^{13}C) in chloroform-d_1 and 3.31 ppm (^{1}H) or 49.1 ppm (^{13}C) in methanol-d_4. Multiplicities of signals are described as follows: s = singlet, bs = broad singlet, d = doublet, t = triplet, q = quartet, quin. = quintuplet, m = multiplet, dd = doublet of doublets. The spectra were analyzed according to first order and coupling constants (*J*) are given in Hertz (Hz). Abbreviations for signal assignment include H_{Ar} = aromatic proton, C_{Ar} = aromatic carbon, C_q = quaternary carbon.

Mass spectroscopy (MS)

Electron ionization (EI, 70 eV) and fast atom bombardment (FAB) mass spectra were measured on a *Finnigan* MAT 90 spectrometer. The molecule fragments are expressed in mass to charge ratio (m/z) and the intensities are given as a percentage relative to the base signal (100%). The molecular ion and protonated molecular ion are referred to as $[M]^+$ and $[M+H]^+$, respectively.

Matrix-assisted laser desorption ionization time of flight (MALDI-TOF) mass spectra of the peptoids were recorded on a *Bruker* Biflex IV spectrometer with a pulsed ultraviolet nitrogen laser (200 μJ at 337 nm) and a time-of-flight mass analyzer with a 125 cm linear flight path. For every spectrum the samples were shot between 100 and 300 times with a repetition rate of 1–3 Hz. The software used for recording and processing the spectra was XACQ Version 4.0.4 and XMASS_TOF Version 5.1.0. The samples were spotted on a *Bruker* Standard stainless steel target 386 spots. As matrix 2,5-dihydroxybenzoic acid (DHB) or a 1:1 mixture of DHB and α-cyano-4-hydroxy-cinnamic acid (CHCA) (Universal MALDI matrix from *Fluka*) as saturated solutions in 50% acetonitrile in water with 0.1% trifluoroacetic acid (TFA) were used.

Infrared spectroscopy (IR)

IR Spectra were registered on a *Bruker* IFS 88. Oils were measured as films between KBr plates. Solid samples were measured with ATR (Attenuated Total Reflection) technique. Wavenumbers (ν) of the absorption bands are given in cm^{-1}. Intensities are described as follows: vs = very strong (0-10% T), s = strong (11-40% T), m = middle (41-70% T), w = weak (71-80% T) and vw = very weak (91-100% T).

High-performance liquid chromatography (HPLC)

Reverse phase analytical HPLC was performed in an *Agilent* Series 1100 equipped with a G1322A-degasser, a G1311A-pump, a G1313A-autosampler, a G1316A oven and a G1315B-diode array detector (DAD). The column was a C18 PerfectSil Target (*MZ Analysentechnik*, 3–5 μm, 4.0 × 250 mm) with a flow rate of 1 mL/min. Purity runs were carried out with a gradient of 5–95% acetonitrile in water with 0.1% TFA over 20 minutes (unless stated otherwise) and the purity was calculated by integration of the signals at 218 nm.

Peptoids were purified in a Vydac 218TP Series (*Grace Davison Discovery Sciences*) C18 column (19 cm × 3 cm) using a *Jasco* HPLC of the LC-NetII/ADC Series equipped with a

MD210 Plus multiwavelength detector, PU-2087 Plus pumps, a CO2060 Plus thermostat and a CHF-122SC fraction collector. An acetonitrile gradient in water with 0.1% TFA was used as eluent. The flow rates varied between 10–15 mL/min. The method (gradient, temperature and flow rate) was adjusted to each sample.

Some peptoid repurifications were completed in a semipreparative HPLC from *Agilent* 1200 Series equipped with a G1322A-degasser, a G1311A-pump, a G1329A-autosampler, a G1316A oven, a G1315D-diode array detector (DAD), and a G1364C fraction collector. A Zorbax 300SB-C8 column (5 µm, 9.4 mm × 250 mm) from *Agilent* and a Lichrosphere C4 column (10 µm, 4 mm × 250 mm) were used for the separations. The flow rates varied between 1–3 mL/min. The method (gradient, temperature and flow rate) was adjusted to each sample.

Lyophilisator

Freeze-drying of aqueous solutions was performed in a lyophilisator from *Christ* model Alpha 1-2 LD plus, equipped with a *Vacuubrand* RZ 2.5 pump.

Optical rotation

Optical rotations were measured on a Perkin Elmer 241 polarimeter and are given as $[\alpha]^{21}_{D} = \alpha/(\beta \cdot d)$ where α is the measured value in °, β is the concentration in g/mL and d is the length of the cuvette in dm. The concentration c is given in g/100 mL.

Circular dichroism (CD)

CD spectra were recorded at 25 °C using a J-815 spectropolarimeter from *Jasco*. The spectra were scanned between 280 and 180 at 0.1 nm intervals. Three repeat scans at a scan rate of 10 nm/min, 8 s response time, and 1 nm bandwidth were averaged for each sample and its respective blank.

Melting points (m.p.)

Melting points were measured on a *Stanford Research System* device, model Opti Melt SRS and are uncorrected.

Microwave (MW)

Microwave reactions were carried out in a single mode *CEM* Discover LabMate microwave operated with *CEM*'s Synergy™ software. This instrument works with a constantly focused power source (0–300W). Irradiation can be adjusted *via* Power- or Temperaturecontrol. The temperature was monitored with an optical fiber (for reactions in open vessel at atmospheric pressure), or an infrared sensor (for reactions in a closed system).

Shaker

For solid-phase reactions at room temperature the resins were shaken on a KS 501 digital circular shaker from *Ika-Labortechnik*.

Analytical scale

Reagents and products were weighted on a *Sartorius* analytical scale models LA310S and BP211D.

6.3 Biological Tests

Cell lines

The human pancreatic cancer cells L3.6pl were kindly provided by C. Bruns (University of Munich, Germany) and were maintained in RPMI medium supplemented with 10% fetal calf serum (FCS). The human tongue squamous cell carcinoma SCC-25 and the human embryonic kidney cells HEK293 were purchased from *ATCC* and were grown in a 1:1 mixture of DMEM (Dulbecco's Modified Eagle Medium) and nutrient mixture F12 (*Life Technologies*) supplemented with 10% FCS and 400 ng/mL hydrocortisone (*Sigma Aldrich*).

Antibodies and other reagents

For Western blot analysis the following antibodies were used: phospho-Erk, phospho-Met (D-26), and Met (25H2) from *Cell Signaling Technology* (Beverly, England), and Erk (K-23) from *Santa Cruz Biotechnology* (Heidelberg, Germany).

The CD44-v6 rat and human peptides have been previously reported.[106] The sequence of the rat 14mer is KEKWFENEWQGKNP and the human 14mer corresponds to KEQWFGNRWHEGYR. The rat and human 5mers are NEWQG and NRWHE, respectively.

Scratch assay

The scratch assays were performed by Dr. Anja von Au in the research group of PD Dr. Véronique Orian-Rousseau in the Institute of Toxicology and Genetics (ITG) in the Karlsruhe Institute of Technology (KIT).

To test the migration potential of several cell lines the cells were seeded in 12-well plates (L3.6pl 2.5 × 10^5 cells/well, Panc1 3.0 × 10^5 cells/well, SCC-25 2.0 × 10^5 cells/well) at 37 °C overnight and the next day serum starved for 24 hours before the assay. In each well a scratch with a pipette tip was performed and the wells were washed once in medium without FCS. The cells were then incubated with the corresponding peptides and peptoids (100 ng/mL or 200 ng/mL) for 10 min at 37 °C before being induced with 10 ng/mL hepatocyte growth factor (HGF) for 24 h. At the time points 0 h and 24 h a picture from each scratch was taken and the areas between the two cell fronts determined with Image J software.

Scatter assay

The scattering assays were performed by Dr. Anja von Au in the research group of PD Dr. Véronique Orian-Rousseau in the Institute of Toxicology and Genetics (ITG) in the Karlsruhe Institute of Technology (KIT).

Scattering of SCC-25 cells was assessed in 24-well plates. Here, 2.0 × 10^4 cells/well were seeded at 37 °C and the next day serum starved for 24 h. Then, the cells were incubated with the corresponding peptides and peptoids (100 ng/mL or 200 ng/mL) for 10 min and induced with 10 ng/mL HGF for 48 h. The human CD44-v6 peptides were used as a positive control. Pictures were taken 48 h after HGF induction.

Western blot

The Western blots were performed by Mark Schmitt in the research group of PD Dr. Véronique Orian-Rousseau in the Institute of Toxicology and Genetics (ITG) in the Karlsruhe Institute of Technology (KIT).

HEK293 cells were serum starved for 24 hours before treatment with the indicated peptide or peptoids (100 ng/mL) concentration for 10 min at 37 °C. The cells were then induced with the growth factor HGF (10 ng/mL) at 37 °C for 5 minutes. Following the induction, the cells were lysed in SDS-sample buffer containing 100 mM dithiothreitol (DTT) and boiled for 5 min at 95 °C. Samples were loaded and blotted against phosphor-Met and Erk. Afterwards the membranes were stripped (62.5 mM Tris, pH 6.8, 2% SDS, 0.8% DTT for 45 min at 55 °C) and blotted against total Met and Erk as loading controls. Staining of the blots was performed by using the enhanced chemiluminescence system from *Thermo Fisher Scientific*.

Animal experiments

The animal experiments were performed by Dr. Anja von Au in the research group of PD Dr. Véronique Orian-Rousseau in the Institute of Toxicology and Genetics (ITG) in the Karlsruhe Institute of Technology (KIT).

Seven-week-old male athymic nude mice were purchased from *Harlan*, housed and maintained under SPF (Specific Pathogen Free) conditions. L3.6pl human pancreatic carcinoma cells were detached by trypsinization and suspended in D-PBS (Dulbecco's Phosphate-Buffer Saline) with a concentration of 2.0×10^7 cells/mL. 50 μL of the cell suspension was injected orthotopically in the head of the pancreas. One week later the mice were randomly distributed in two groups of 5 mice and either CD44-v6 peptoid **38** (for the treatment group) or control peptoid **107** (for the control group) were injected intraperitoneally. The injection was repeated three times per week for 21 days. Three days after the last treatment the animals were killed and the tumor and liver were prepared. After determination of size and weight the organs were fixed in 10 % neutral buffered formaldehyde and for a part of the primary tumor a zinc fixation was performed to allow blood vessel determination by CD31 staining.

Histology

Paraffin embedded liver, and primary tumors were stained with hematoxylin and eosin stain (H&E) for histological analysis. The whole liver tissue was examined for the presence and the extension of metastasis formation.

Immunhistological analysis

Paraffin sections were deparaffinized and rehydrated. For blood vessel staining a CD31 antibody was used. The unmasking was achieved by boiling the slides in 1 mM EDTA pH 8.0.

Immunostaining and microscopy.

Immunofluorescence experiments were performed by Dr. Antje Neeb in the research group of Prof. Cato in the Institute of Toxicology and Genetics (ITG) in the Karlsruhe Institute of Technology (KIT).

For immunofluorescence 2×10^4 22Rv.1 prostate cancer cells were seeded in 1 mL RPMI

10% FCS on a sterile glass cover slip in a 24 well cell culture dish. 24 h after seeding, the Bag-1- or control peptoids were added (1 μM; 0.1% DMSO) and incubated for 20 h. Cells were washed with D-PBS and fixed with 4% Paraformaldehyd (PFA) in D-PBS (5 min), After fixation, PFA was removed, cells washed twice with PBS and incubated with permeabilisation solution (0,2% Triton X-100 in PBS) for 5 min at room temperature. Then cells were washed twice with PBS, incubated with blocking solution (4% goat serum in PBS 1X) for 5 min at room temperature and subsequently incubated with primary antibody (rabbit anti GRP75; *Santa Cruz* H155 or rabbit anti GRP78; *Abcam* ab21685) diluted 1:100 in blocking buffer for 20 min at room temperature. The cells were then washed twice with PBS and secondary antibody (Alexa 488 or Alexa 546 goat anti rabbit; *Cell signaling*) diluted 1:1000 in blocking buffer was applied for 20 min at room temperature. Cell nuclei were counterstained using Draq5 (1:1000 in blocking buffer, 20 min, r.t.). Finally, the cells were washed twice with PBS 1X and once with water, dried and mounted on a glass slide using polyvinylalcohol (PVA) as mounting medium. Samples were then analyzed with a Zeiss LSM 200 confocal microscope using LSM software.

FACS binding assay.

The FACS binding assay was performed by Dr. Antje Neeb in the research group of Prof. Cato in the Institute of Toxicology and Genetics (ITG) in the Karlsruhe Institute of Technology (KIT).

BPH-1 benign prostate and 22Rv.1 prostate cancer cells were trypsinised, harvested in medium, washed once with PBS and 2×10^5 cells were incubated with 10 nM-10 μM FITC-labeled Bag-1-peptoid diluted in 100 μL PBS/10%FCS for 1 h at 4 °C. After incubation, cells were washed three times with PBS/10%FCS and analysed on a FACScan using the CelQuest Pro software (BD Bioscience).

6.4 General Procedures

GP1: Swelling of Rink-Amide resin and Fmoc deprotection

1 equiv. of dry Rink-Amide resin is swollen in a glass or plastic fritted-syringe with twice its volume of DMF for 2 h. Then the solvent is removed.

For Fmoc deprotection the resin is treated with a solution of 20% piperidine in DMF (approx. 1 mL/100 mg resin) for 5 min at room temperature. The solvent is filtered off and the procedure is repeated two more times. Afterwards the resin is washed with DMF (× 4).

GP2: Microwave-assisted solid-phase peptoid synthesis *via* monomer method

A 0.1–0.2 M solution of Fmoc-protected monomer (3.00 equiv.), HOBt (3.00 equiv.) and DIC (3.00 equiv.) in peptide-synthesis grade DMF is added to the resin (1.00 equiv.). The mixture is stirred at 60 °C for 30 min under microwave irradiation (max. power: 20 W, open vessel, optical fiber temperature control). The reaction solution is filtered out and the coupling step is repeated (double coupling). Afterwards the resin is washed with DMF (× 4).

For Fmoc deprotection the resin is treated with a solution of 20% piperidine in DMF (approx. 1 mL/100 mg resin) for 5 min at room temperature. The solvent is filtered off and the procedure is repeated two more times. Afterwards the resin is washed with DMF (× 4).

GP3: Microwave-assisted solid-phase peptoid synthesis *via* submonomer method with bromoacetic acid

A 1 M solution of bromoacetic acid (7.90–9.00equiv.) and DIC (7.00–7.90 equiv.) in peptide-synthesis grade DMF is added to the resin (1.00 equiv.). The mixture is stirred at 35 °C under microwave irradiation (max. power: 40 W, open vessel, optical fiber temperature control) for 1 min. Then, the resin is washed with DMF (× 4).

Next, a 1 M solution of the desired amine (7.90–10.0 equiv.) in peptide-synthesis grade DMF or N-methyl-2-pyrrolidone (NMP) is added to the resin (1.00 equiv.). The mixture is stirred at 60 °C under microwave irradiation (max. power: 40 W, open vessel, optical fiber temperature control) for 30 or 45 min. Finally, the resin is washed with DMF (× 4).

GP4: Microwave-assisted solid-phase peptoid synthesis *via* submonomer method with chloroacetic acid

A 1 M solution of chloroacetic acid (7.90–9.00 equiv.) and DIC (7.90–9.00 equiv.) in peptide-synthesis grade DMF is added to the resin (1.00 equiv.). The mixture is stirred at 35 °C under microwave irradiation (max. power: 40 W, open vessel, optical fiber temperature control) for 10 min. Then, the resin is washed with DMF (× 4).

Next, a 1 M solution of the desired amine (9.00–10.0 equiv.) in peptide-synthesis grade DMF or NMP is added to the resin (1.00 equiv.). The mixture is stirred at 60 °C under microwave irradiation (max. power: 40 W, open vessel, optical fiber temperature control) for 45 min. Finally, the resin is washed with DMF (× 4).

GP5: Room temperature solid-phase peptoid synthesis *via* monomer method

A 0.1–0.5 M solution of Fmoc-protected monomer (3.00 equiv.), HOBt (3.00 equiv.) and DIC (3.00 equiv.) in peptide-synthesis grade DMF is added to the resin (1.00 equiv.) and the mixture is shaken at room temperature for the specified amount of time. The reaction solution is filtered out and the coupling step is repeated (double coupling). Finally, the resin is washed with DMF (× 4).

Alternatively a 0.1 M solution of Fmoc-protected monomer (5.00 equiv.), HOBt (5.00 equiv.) and DIC (5.00 equiv.) in peptide-synthesis grade DMF is added to the resin and the mixture is shaken at room temperature overnight.

For Fmoc deprotection the resin is treated with a solution of 20% piperidine in DMF (approx. 1 mL/100 mg resin) for 5 min at room temperature. The solvent is filtered off and the procedure is repeated two more times. Afterwards the resin is washed with DMF (× 4).

GP6: Room temperature solid-phase peptoid synthesis *via* submonomer method with bromoacetic acid

A 1 M solution of bromoacetic acid (7.90–9.00 equiv.) and DIC (7.00–7.90 equiv.) in peptide-synthesis grade DMF is added to the resin (1.00 equiv.). The mixture is shaken at room temperature for the indicated amount of time. Then, the resin is washed with DMF (× 4).

Next, a 1 M solution of the desired amine (7.90–10.0 equiv.) in peptide-synthesis grade DMF or NMP is added to the resin. The mixture is shaken at room temperature for the specified amount of time. Finally, the resin is washed with DMF (× 4).

GP7: Room temperature solid-phase peptoid synthesis *via* the submonomer method with chloroacetic acid

A 0.5–1 M solution of chloroacetic acid (9.00 equiv.) and DIC (7.00–9.00 equiv.) in peptide-synthesis grade DMF is added to the resin. The mixture is shaken at room temperature for the specified amount of time. Then, the resin is washed with DMF (× 4).

Next, a 1 M solution of the desired amine (8.30–9.00 equiv.) in peptide-synthesis grade DMF or NMP is added to the resin. The mixture is shaken at room temperature for the specified amount of time. Finally, the resin is washed with DMF (× 4).

GP8: Peptoid synthesis on 2-Chlorotrityl chloride resin

1.00 equiv. of dry 2-Chlorotrityl chloride resin is swollen in a plastic fritted-syringe with twice its volume of dichloromethane for 30 min. After removal of the solvent a 0.5 M solution of bromoacetic acid (submonomer method) or Fmoc-protected monomer (monomer method) (1.30 equiv.) and DIPEA (5.20 equiv.) in dichloromethane is added to the resin. The mixture is shaken at room temperature for 45 min–1.5 h. Finally, the resin is washed with dichloromethane (× 3) and DMF (× 3). Amine substitution and following couplings take place as described in GP5-GP7.

GP9: Guanidination on solid-support

The procedure for guanidination on solid-phase was adapted from Eggleston.[136] *N*, *N*'-di-Boc-1*H*-pyrazole-1-carboxamidine or 1*H*-pyrazole-1-carboxamidine monohydrochloride (5.00 equiv. per amine side-chain) and DIPEA (7.50 or 15.0 equiv. per amine side-chain) are dissolved in 1 mL peptide-synthesis grade DMF and stirred overnight at room temperature. Finally, the resin is washed with DMF (× 4).

GP10: Guanidination in solution

In a microwave vessel 1.00 equiv. of the lyophilized peptoid and 1*H*-pyrazole-1-carboxamidine monohydrochloride (3.00 equiv. per amine side-chain) are dissolved in 1–2 mL dimethylacetamide (DMA). DIPEA (9.00 equiv. per amine side-chain) is added, the vessel is sealed and the mixture is stirred at 60 °C under microwave irradiation for 2 h (max. power: 60 W, IR temperature control). Finally, the mixture is diluted with 5–10 mL water and lyophilized.

GP11: Peptoid capping on solid-phase

Acetic acid anhydride (10.0 equiv.) and DIPEA (15.0 equiv.) dissolved in 0.5–1 mL DMF are added to the resin. The mixture is stirred/ shaken at room temperature for the specified amount of time. Finally, the resin is washed with DMF (× 4).

GP12: Cleavage from Rink-Amide Resin

A 95% TFA solution in dichloromethane is added to the resin (approx. 1 mL/100 mg resin) and let sit or shaken at room temperature for 1.5–4 h. The solution is collected, new cleavage mixture is added to the resin and let sit or shaken for 1.5–4 h more. Alternatively the cleavage mixture is added to the resin and let sit overnight. After filtering out the solution, the resin is washed with the same volume of dichloromethane (× 3) and methanol (× 3). The cleavage solutions and washes are combined and the solvent is removed under reduced pressure. The residue is taken up in approx. 5 mL water and lyophilized.

GP13: Cleavage from Chlorotrityl resin

A 20% 1,1,1,3,3,3-hexafluoroisopropanol (HFIP) solution in dichloromethane is added to the resin (approx. 1 mL/100 mg resin) and shaken at room temperature for 1–2 h. The solution is collected, new cleavage mixture is added to the resin and shaken for 1–2 h more. After filtering out the new solution, the resin is washed with the same volume of dichloromethane (× 3) and methanol (× 3). The cleavage solutions and washes are combined and the solvent is removed under reduced pressure.

GP14: Peptoid cyclization

Peptoid cyclization was performed according to Kirshenbaum.[54] The linear peptoid is dissolved in dry DMF forming a 2 μM solution and argon is bubbled through for 5 min. Then DIPEA (6.00 equiv.) and PyBOP (3.00 equiv.) are added and the reaction mixture is stirred at room temperature for 1 h. Then the solvent is evaporated under reduced pressure. The residue is taken up in approx. 5 mL water and lyophilized.

6.5 Synthesis and Characterization of Compounds

Submonomer syntheses

***N*-(2-(1*H*-indol-3-yl)ethyl)-2,2,2-trifluoroacetamide (47)**

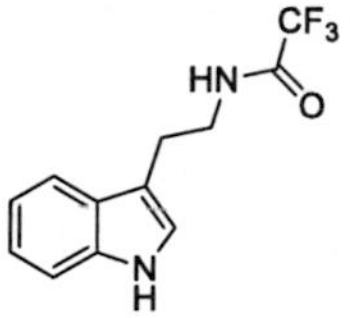

CF_3 HN O N H

Compound **47** was synthesized as described by Zuckermann.[114] To a solution of 3.01 g tryptamine (**39**, 18.8 mmol, 1.00 equiv.) in 60 mL pyridine/dichloromethane (1:1) mixture, 3.94 mL trifluoroacetic anhydride (**46**, 5.92 g, 28.2 mmol, 1.50 equiv.) were added slowly at 0 °C. The reaction was stirred at room temperature for 3.5 h. Then the crude was extracted with water (3 × 30 mL), and saturated citric acid solution (3 × 30 mL) and the organic layer was dried over anhydrous sodium sulfate. After evaporation of the solvent the product was purified by column chromatography over silica gel (100% dichloromethane) affording 3.36 g of the product (13.1 mmol, 70% yield) as a colorless oil – ^{1}H-NMR (250 MHz, $CDCl_3$): δ/ppm = 3.07 (t, *J* = 6.6 Hz, 2 H, CH_2C_{Ar}), 3.71 (q, *J* = 6.6 Hz, 2 H, CH_2NH), 7.06 (d, *J* = 2.4 Hz, 1 H, C*H*N), 7.16 (m, 1 H, H_{Ar}), 7.25 (m, 1 H, H_{Ar}), 7.41 (d, *J* = 8.1 Hz, 1 H, H_{Ar}), 7.60 (d, *J* = 7.6 Hz, 1 H, H_{Ar}), 8.08 (bs, 1 H, NH_{Ar}). Analytics are consistent with those reported in the literature.[137]

***tert*-butyl 3-(2-(2,2,2-trifluoroacetamido)ethyl)-1*H*-indole-1-carboxylate (49)**

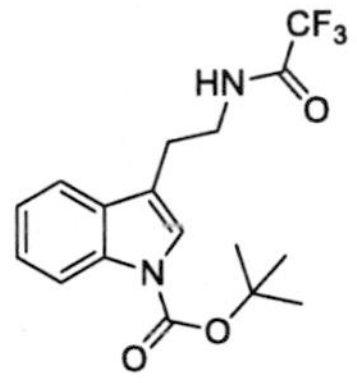

Compound **49** was synthesized as described by Zuckermann.[114] To a solution of starting material **47** (3.34 g, 13.4 mmol, 1.00 equiv.) in 40 mL THF, 3.35 mL Boc-anhydride (**48**, 3.42 g, 15.7 mmol, 1.20 equiv.) and 80 mg DMAP (0.652 mmol, 0.05 equiv.) were added and the reaction mixture was stirred at 40 °C for 2.5 h. After evaporation of the solvent, the residue was dissolved in ethyl acetate (40 mL) washed with brine (3 × 40 mL), and the organic layer was dried over anhydrous sodium sulfate. Column chromatography over silica gel (cyclohexane/ ethyl acetate 4:1) afforded 3.84 g of compound **49** (10.8 mmol, 83% yield) as a colorless oil. ^{1}H-NMR (250 MHz, $CDCl_3$): δ/ppm = 1.67 (s, 9 H, $C(CH_3)_3$), 3.00 (td, *J* = 6.8 Hz, *J* = 0.8 Hz, 2 H, CH_2C_q), 3.70 (q, *J* = 6.8 Hz, 2 H, CH_2NH), 6.36 (bs, 1 H, N*H*), 7.27 (m, 1 H, H_{Ar}), 7.35 (m, 1 H, H_{Ar}), 7.44 (s, 1 H, C*H*N), 7.50–7.55 (m, 1 H, H_{Ar}), 8.15 (d, *J* = 8.0 Hz, 1 H, H_{Ar}). Analytics are consistent with those reported in the literature.[137]

***tert*-butyl 3-(2-aminoethyl)-1*H*-indole-1-carboxylate (40)**

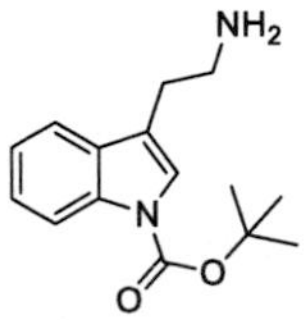

Compound **40** was synthesized as described by Zuckermann.[114] Starting material **49** (3.84 g, 10.8 mmol, 1.00 equiv.) was dissolved in 384 mL of 5% potassium carbonate solution in a 70% methanol solution in water. The reaction was stirred overnight at room temperature.

Then the crude was concentrated under reduced pressure and extracted with ethyl acetate (3 × 100 mL). The combined organic layers were washed with water (1 × 250 mL) and brine (2 × 250 mL), and dried over sodium sulfate. Column chromatography over silica gel (4% methanol in dichloromethane + 1% TEA) afforded 2.15 g of product **40** (8.26 mmol, 77% yield) as a colorless oil. – ^{1}H-NMR (250 MHz, $CDCl_3$): δ/ppm = 1.66 (s, 9 H, $C(CH_3)_3$), 2.91 (t, J = 6.8 Hz, 2 H, CH_2), 3.70 (t, J = 6.8 Hz, 2 H, CH_2), 7.20–7.26 (m, 1 H, H_{Ar}), 7.28–7.35 (m, 1 H, H_{Ar}), 7.45 (s, 1 H, C*H*N), 7.52–7.56 (m, 1 H, H_{Ar}), 8.10–8.14 (m, 1 H, H_{Ar}). Analytics are consistent with those reported in the literature.[137]

N,N'-Di-Boc-1H-Diaminopentan-1-carboxamidine (42)

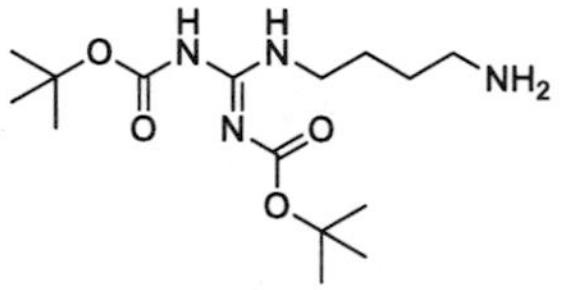

A solution of 2.00 g *N,N'*-Di-Boc-1*H*-Pyrazole-1-carboxamidine (**50**, 6.44 mmol, 1.00 equiv.) in 22 mL dimethylformamide was added slowly at room temperature under stirring to a solution of 1.94 mL 1,4-Diaminobutane (**51**, 1.70 g, 19.3 mmol, 3.00 equiv.) in 120 mL dimethylformamide. The reaction was stirred overnight at room temperature. The solvent was evaporated under reduced pressure and the product was purified by column chromatography over silica gel (ethyl acetate/methanol/triethylamine 7:1:1) affording 1.83 g (5.54 mmol, 86% yield) of a colorless oil. – ^{1}H-NMR (300 MHz, MeOD): δ/ppm= 1.47 (s, 9 H, $C(CH_3)_3$), 1.53 (s, 9 H, $C(CH_3)_3$), 1.40–1.65 (m, 4 H, $CH_2(CH_2)_2CH_2$), 2.67 (t, J = 6.9 Hz, 2 H, CH_2NH_2), 3.37 (t, J = 6.9 Hz, 2 H, CH_2NHCN) – ^{13}C-NMR (100 MHz, MeOD): δ/ppm = 27.43 (CH_2), 28.27 (CH_3), 28.61 (CH_3), 30.12 (CH_2), 41.42 (CH_2), 41.83 (CH_2), 80.43 ($C(CH_3)_3$), 84.52 ($C(CH_3)_3$), 154.26 (*C*O), 157.68 (*C*O), 164.61 (*C*N) – IR (ATR): ν(cm^{-1}) = 3333 (w), 2978 (w), 2932 (w), 1715 (m), 1610 (m), 1568 (m), 1413 (m), 1365 (m), 1323 (m), 1251 (m), 1227 (m), 1126 (s), 1050 (m), 1023 (m), 877 (w), 851 (w), 807 (m), 761 (m), 616 (w) – MS (EI, 70eV): m/z (%): 330 (61) $[M]^+$, 274 (37) $[M+H–tBu]^+$, 257 (26) $[M–tBuO]^+$ 230 (28) $[M+H–Boc]^+$, 218 (23) $[M+2H–2tBu]^+$, 201 (28) $[M+H–tBu–tBuO]^+$, 174 (33) $[M+2H–Boc–tBu]^+$, 157 (62) $[M+H–Boc–tBuO]^+$, 130 (49) $[M+2H–2xBoc]^+$, 87 (100) $[NH_2(CH_2)_4NH]^+$, 57 (62) $[tBu]^+$. Analytics are consistent with those reported in the literature.[138]

4-(((Benzyloxy)carbonyl)amino)butanoic acid (53)

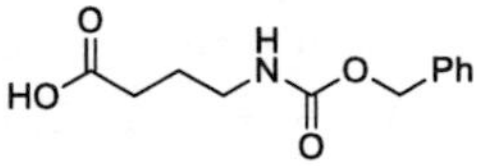

To a solution of γ-aminobutanoic acid (**52**, 5.00 g, 48.5 mmol, 1.00 equiv.) in 25 mL 2 M NaOH, a suspension of 6.10 mL benzyl chloroformate (4.20 g, 52.4 mmol, 1.08 equiv.) and 25 mL 2 M NaOH was slowly added at 0 °C. The reaction was stirred at 0°C for 40 min and then at room temperature overnight. The crude was washed with diethyl ether (3 × 50 mL), acidified to pH = 3 with conc. HCl and cooled in an ice bath. The white precipitate was filtered, washed with cold 0.1 M HCl aq. solution and dried to yield 9.80 g (41.3 mmol, 85% yield) of a white solid. – m.p. = 69 °C. – ^{1}H-NMR (300 MHz, MeOD): δ/ppm = 1.78 (quin, J = 7.1 Hz, 2 H, $CH_2\mathit{CH_2}CH_2$), 2.31 (t, J = 7.1 Hz, 2 H, $\mathit{CH_2}$COOH), 3.16 (t, J = 7.1 Hz, 2 H, $\mathit{CH_2}$NH), 5.06 (*s*, 2 H, $\mathit{CH_2}$-Ph), 7.22–7.35 (m, 5 H, H_{Ar}). – ^{13}C-NMR (75 MHz, MeOD): δ/ppm = 25.1 (CH_2), 31.5 (CH_2), 40.2 (CH_2), 66.7 (CH_2), 128.07 (C_{Ar}), 128.11 (C_{Ar}), 128.5 (C_{Ar}), 136.4 (C_q, C_{Ar}), 156.7 (C_q, CO_2H), 178.4 (C_q, *C*ONH). – IR (KBr): ν (cm^{-1}) = 3335 (w), 2950 (w), 1712 (s), 1537 (m), 1454 (m), 1360 (w), 1261 (m), 1025 (w), 739 (w), 698 (w). – MS (EI, 70 eV), m/z (%): 237 (71) $[M]^+$, 108 (100) $[BnOH]^+$, 91 (72) $[PhCH_2]^+$. – HRMS (EI, 70 eV): $C_{12}H_{15}NO_4$, calc.: 237.1001; found: 237.0999.

***tert*-Butyl 4-(((benzyloxy)carbonyl)amino)butanoate (54)**

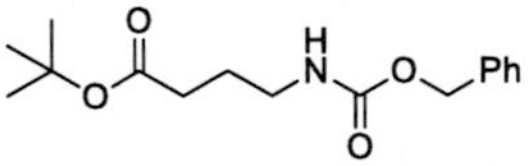

Procedure a):[115] Carboxylic acid **53** (2.44 g, 10.3 mmol, 1.00 equiv.), *tert*-butanol (2.95 mL, 2.29 g, 30.9 mmol, 3.00 equiv.) and DMAP (126 mg, 1.03 mmol, 0.10 equiv.) were dissolved in 50 mL dry dichloromethane. DCC (2.44 g, 11.8 mmol, 1.15 equiv.) was added to the mixture and the reaction was stirred at room temperature overnight. After filtering off the precipitated dicyclohexylurea, the filtrates were washed with water (2 × 50 mL), 1 M HCl (2 × 50 mL), 5% $NaHCO_3$ (2 × 50 mL) and brine (2 × 50 mL). The combined organic layers were dried over magnesium sulfate and the solvent was evaporated. Purification *via* column chromatography over silica gel (cyclohexane/ ethyl acetate 2:1) afforded 1.59 g (5.44 mmol, 53% yield) of the product as a yellow oil.

Procedure b):[116] Carboxylic acid **53** (239 mg, 1.01 mmol, 1.00 equiv.) was dissolved in 5 mL dry toluene. Under argon 0.14 mL dry triethylamine (102 mg, 1.01 mmol, 1.00 equiv.) and 0.16 mL 2,4,6-trichlorbenzoyl chloride (250 mg, 1.02 mmol, 1.01 equiv.) were added to the mixture and stirred at room temperature for 1 h. Then a solution of dry *tert*-butanol (0.17 mL, 132 mg, 1.78 mmol, 1.76 equiv.) and DMAP (222 mg, 1.82 mmol, 1.80 equiv.) in 3 mL toluene was added. The reaction mixture was stirred overnight at room temperature. Then the crude was diluted with 10 mL diethyl ether and washed with saturated aqueous $NaHCO_3$ (3 × 10 mL). The washes were extracted again with diethyl ether (10 mL) and the combined organic layers were dried over magnesium sulfate. Evaporation of the solvent and purification by flash column chromatography over silica gel (cyclohexane/ethyl acetate 3:1) afforded 188 mg (0.641 mmol, 63% yield) of the product as a yellow oil. – ^{1}H-NMR (300 MHz, $CDCl_3$): δ/ppm = 1.43 (s, 9 H, $C(CH_3)_3$), 1.73–1.87 (quin, J = 7.1 Hz, 2 H, $CH_2CH_2CH_2$), 2.27 (t, J = 7.1 Hz, 2 H, CH_2CO), 3.23 (apparent q, J = 6.5 Hz, 2 H, CH_2NH), 4.86 (bs, 1 H, NH), 5.09 (s, 2 H, CH_2Ph), 7.31–7.84 (m, 5 H, H_{Ar}). Analytics are consistent with the literature.[116]

tert-Butyl 4-aminobutanoate (43)

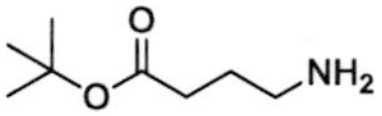

Amide **54** (3.21 g, 11.0 mmol, 1.00 equiv.) was dissolved in 50 mL methanol and 10 mass% of the catalyst (5% Pd/C) was added. The obtained suspension was treated with H_2 at room temperature for 4.5 h. After completion of the reaction, the catalyst was removed by filtration over Celite® and the solvent was evaporated yielding 1.67 g (10.5 mmol, 95% yield) of a brown oil. – ^{1}H-NMR (250 MHz, $CDCl_3$): δ/ppm = 1.35 (s, 9 H, $C(CH_3)_3$), 1.58–1.70 (m, 2 H, $CH_2CH_2CH_2$), 1.73 (bs, 2 H, NH_2) 2.18 (t, J = 7.4 Hz, 2 H, CH_2CO), 2.63 (t, J = 7.0 Hz, 2 H, CH_2N). Analytics are consistent with the literature.[116]

3-((_tert_-butoxycarbonyl)amino)propanoic acid (56)

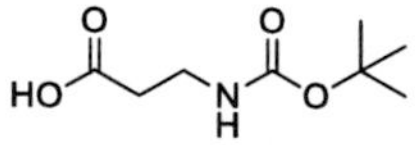

Compound **56** was synthesized following a procedure adapted from Keillor.[118] To a solution of 2.29 g β-alanine (**55**, 25.7 mmol, 1.00 equiv.) in 50 mL methanol, 11.9 mL Boc anhydride (**48**, 12.1 g, 54.0 mmol, 2.10 equiv.) and 21.4 mL triethylamine (15.6 g, 154 mmol, 6.00 equiv.) were added and the reaction mixture was stirred at room temperature for two days. After evaporation of the solvent, the crude was dissolved in 50 mL of saturated $NaHCO_3$ solution and extracted with cyclohexane (3 × 50 mL). The aqueous phase was acidified to pH = 2 with conc. HCl and extracted with ethyl acetate (3 × 50 mL). The combined organic layers were washed with brine and dried over sodium sulfate. Evaporation of the solvent afforded 4.40 g (23.2 mmol, 90% yield) of the product **56** as a white solid. – ^{1}H-NMR (250 MHz, MeOD): δ/ppm = 1.43 (s, 9 H, C(CH_3)$_3$), 2.47 (t, J = 6.8 Hz, 2 H, CH_2CO), 3.29–3.33 (m, 2 H, CH_2N). Characterization of this molecule has been previously described and analytics were consistent with the literature.[139]

4-nitrophenyl 3-((*tert*-butoxycarbonyl)amino)propanoate (58)

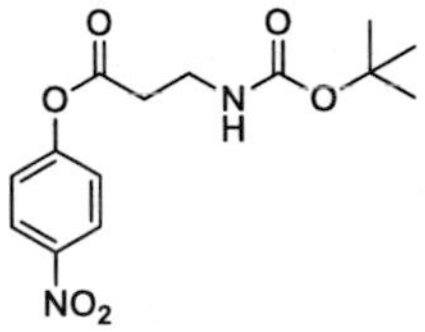

Compound **58** was synthesized following a procedure adapted from Keillor.[118] To a solution of carboxylic acid **56** (4.40 g, 23.2 mmol, 1.00 equiv.) in 30 mL acetonitrile, *p*-nitrophenyl chloroformate (**57**, 5.31 g, 25.6 mmol, 1.10 equiv.), dimethylaminopyridine (DMAP, 280 mg, 2.32 mmol, 0.100 equiv.) and triethylamine (4.0 mL, 2.93 g, 28.1 mmol, 1.10 equiv.) were added. The reaction mixture was stirred overnight at room temperature. After evaporation of the solvent under reduced pressure the residue was taken up in ethyl acetate and precipitated triethylamine hydrochloride was filtered off. The product was purified by column chromatography over silica gel (cyclohexane/ ethyl acetate 1:1) obtaining 5.55 g of a mixture containing 90% product **58** (4.99 g, 16.1 mmol, 69% yield) and 10% p-nitrophenol, which was used in the next step without further purification. – ^{1}H-NMR (250 MHz, MeOD): δ/ppm = 1.44 (s, 9 H, C(CH_3)$_3$), 2.80 (t, J = 6.5 Hz, 2 H, CH_2CO), 3.44 (t, J = 6.5 Hz, 2 H, CH_2N), 7.39 (d, J = 9.0 Hz, 2 H, CH_{Ar}CO), 8.28 (d, J = 9.0 Hz, 2 H, CH_{Ar}CNO$_2$).

***tert*-Butyl (3-amino-3-oxopropyl)carbamate (59)**

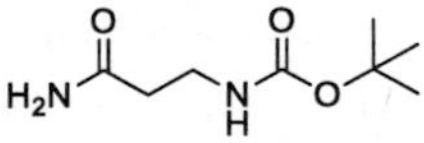

To a solution of compound **58** (4.99 g, 16.1 mmol, 1.00 equiv.) in 50 mL methanol, 7.7 mL of a 7 N solution of NH_3 in methanol (53.9 mmol, 3.35 equiv.) were added at 0 °C and the reaction mixture was stirred at room temperature. TLC showed reaction completion after 4 h. Then, the solvent was evaporated and the product was purified by column chromatography over silica gel (ethyl acetate + 1% AcOH) obtaining 2.00 g of the product (**59**, 10.6 mmol, 66% yield) as a white solid. – ^{1}H-NMR (250 MHz, MeOD): δ/ppm = 1.43 (s, 9 H, C(CH_3)$_3$), 2.38 (t, J = 6.9 Hz, 2 H, CH_2CO), 3.27–3.32 (m, 2 H, CH_2N). Characterization of this molecule has been previously described and analytics were consistent with the literature.[140]

3-Aminopropanamide hydrochloride (44·HCl)

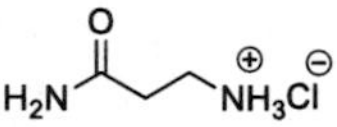

917 mg of compound **59** (4.87 mmol, 1.00 equiv.) were dissolved in 20 mL 2 N HCl and stirred overnight at room temperature. Coevaporation of the solvent with toluene afforded the product (**44·HCl**) in quantitative yield. – ^{1}H-NMR (250 MHz, MeOD): δ/ppm = 2.72 (t, J = 6.5 Hz, 2 H, CH_2CO), 3.18 (t, J = 6.5 Hz, 2 H, CH_2NH_2).

4-Nitrophenyl 4-(((benzyloxy)carbonyl)amino)butanoate (64)

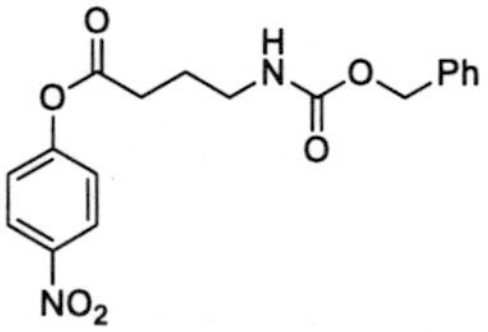

To a solution of carboxylic acid **53** (9.80 g, 41.3 mmol, 1.00 equiv.) in 500 mL ethyl acetate, *p*-nitrophenyl chloroformate (**57**, 9.20 g, 45.5 mmol, 1.10 equiv.), DMAP (0.500 g, 4.13 mmol, 0.100 equiv.) and triethylamine (6.30 mL, 4.60 g, 45.5 mmol, 1.10 equiv.) were added. The reaction mixture was stirred at room temperature for 3 h. After filtering off the precipitated triethylamine hydrochloride, the solution was concentrated under reduced pressure and washed several times with a total of 3 L half-saturated $NaHCO_3$ solution. The organic phase was dried over anhydrous sodium sulfate and the solvent was evaporated under reduced pressure yielding 12.1 g (**64**, 33.8 mmol, 88% yield) of a white solid. - ^{1}H-NMR (300 MHz, $CDCl_3$): δ/ppm = 1.98 (quin, J = 7.1 Hz, 2 H, $CH_2CH_2CH_2$), 2.67 (t, J = 7.1 Hz, 2 H, COCH_2), 3.32–3.38 (m, 2 H, CH_2NH), 5.11 (s, 2 H, CH_2Ph), 7.32–7.37 (m, 5 H, H_{Ar}),

8.16–8.36 (m, 4 H, H_{Ar}). ^{13}C-NMR (75 MHz, $CDCl_3$): δ/ppm = 25.2 (CH_2), 31.6 (CH_2), 40.3 (CH_2), 67.0 (CH_2), 122.6 (C_{Ar}), 125.4 (C_{Ar}), 128.4 (C_{Ar}), 128.7 (C_{Ar}), 136.5 (C_q, $C_{Ar}CH_2$), 145.5 (C_q, $C_{Ar}NO_2$), 155.5 (C_q, $C_{Ar}O$), 156.7 (C_q, OCONH), 171.0 (C_q, COOAr). – IR (ATR): ν (cm^{-1}) = 3313.3 (w), 3074 (w), 2955 (w), 1750 (m), 1686 (s), 1614 (w), 1592 (m), 1552 (m), 1517 (s), 1488 (m), 1447 (m), 1417 (w), 1346 (s), 1314 (m), 1291 (m), 1264 (s), 1210 (s), 1128 (s), 1099 (s), 1018 (m), 944 (m), 912 (m), 884 (w), 853 (m), 832 (w), 755 (m), 695 (m), 640 (m), 581 (m), 550 (w), 498 (m), 434 (w). – MS (EI, 70 eV), m/z (%): 358 (0.2) $[M+H]^+$, 220 (17) $[M–NO_2PhO]^+$, 139 (7) $[NO_2PhOH]^+$, 107 (9) $[BnO]^+$, 91 (100) $[PhCH_2]^+$. – HRMS (EI, 70 eV): $C_{18}H_{18}N_2O_6$, calc.: 358.1159; found: 358.1159.

Benzyl (4-amino-4-oxobutyl)carbamate (65)

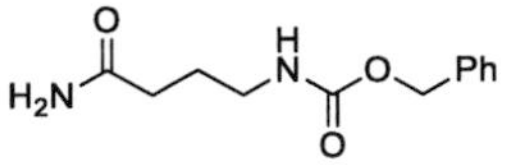

To a solution of compound **64** (12.0 g, 33.8 mmol, 1.00 equiv.) in 5 mL methanol, 5 mL of a 7 N solution NH_3/MeOH were added and the reaction mixture was stirred at room temperature for 3 h. After removal of the solvent the product was purified by column chromatography (gradient 5–20% methanol in ethyl acetate) obtaining 7.32 g (**65**, 31.0 mmol, 92% yield) of a white solid. – m.p. – 127–130 °C – ^{1}H-NMR (300 MHz, MeOD): δ/ppm = 1.78 (quin, J = 6.8 Hz, 2 H, $CH_2CH_2CH_2$), 2.23 (t, J = 6.8 Hz, 2 H, CH_2NH), 3.15 (t, J = 6.8 Hz, 2 H, $COCH_2$), 5.06 (s, 2 H, CH_2Ph), 7.28-7.35 (m, 5 H, H_{Ar}). – ^{13}C-NMR (75 MHz, MeOD): δ/ppm= 27.1 (CH_2), 30.21 (CH_2), 41.3 (CH_2), 67.4 (CH_2), 128.8 (C_{Ar}), 128.96 (C_{Ar}), 129.5 (C_{Ar}), 158 (CO), 164.3 (CO). – IR (ATR): ν (cm^{-1}) = 2966 (w), 2537 (w), 2481 (w), 2379 (w), 2334 (w), 1678 (m), 1624 (m), 1576 (w), 1427 (m), 1383 (w), 1359 (m), 1340 (m), 1303 (w), 1279 (w), 1197 (w), 1150 (m), 1060 (w), 1004 (w), 948 (w), 776 (w), 741 (w), 728 (w), 691 (w), 647 (w), 599 (w), 559 (vw), 517 (w). – MS (EI, 70 eV), m/z (%): 237 (5.3) $[M+H]^+$, 129 (18) $[M-BnO]^+$, 107 (18) $[BnO]^+$, 91 (100) $[CH_2Ph]^+$. – HRMS (EI, 70 eV) $C_{12}H_{16}N_2O_3$, calc. 236.1161; found 236.1159.

4-Aminobutanamide (45)

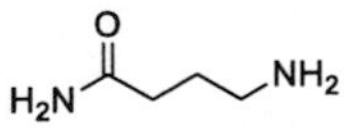

7.32 g of amide **45** (31.0 mmol, 1.00 equiv.) were dissolved in 140 mL methanol and 10 mass% of the catalyst (5% Pd/C) were added. The obtained suspension was treated with H_2 at room temperature overnight. After completion of the reaction, the catalyst was removed by filtration over Celite® and the solvent was evaporated yielding 3.14 g (**45**, 30.8 mmol, 99% yield) of a viscous solid. – ^{1}H-NMR (300 MHz, MeOD): δ/ppm = 1.52 (quin, J = 7.5 Hz, 2 H, CH_2*CH_2*CH_2), 2.01 (t, J = 7.5 Hz, 2 H, CO*CH_2*), 2.45 (t, J = 7.5 Hz, 2 H, *CH_2*NH_2), 6.05 (bs, 2 H, CH_2*NH_2*), 7.06 (bs, 2 H, *NH_2*CO). – ^{13}C-NMR (75 MHz, MeOD): δ/ppm = 29.8 (*CH_2*), 33.9 (*CH_2*), 42.0 (*CH_2*), 178.7 (*CO*). – IR (ATR): ν (cm^{-1}) = 3339 (m), 3175 (m), 2944 (m), 2870 (w), 1643 (m), 1465 (w), 1415 (m), 1372 (m), 1320 (m), 1264 (w), 1238 (w), 1149 (w), 1103 (w), 1058 (w), 966 (w), 914 (m), 845 (w), 742 (m), 637 (m), 519 (w). – MS (EI, 70 eV), m/z (%): 103 (17) $[M+H]^+$, 59 (55) $[M-H_2NCO]^+$, 44 (100) $[M-(CH_2)_3NH_2]^+$. – HRMS (EI, 70 eV): $C_4H_{10}N_2O$, calc. 102.0793; found 102.0796.

3-Aminopropanamide (44)

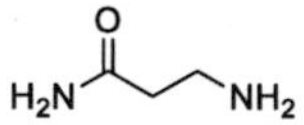

303 mg cyanoacetamide (**66**, 3.60 mmol, 1.00 equiv.) and PtO_2 (40.0 mg, 4.68 mmol, 1.30 equiv.) were suspended in acetic acid and treated with H_2 at room temperature overnight. The catalyst was filtered off over Celite® and the solvent was removed under reduced pressure by coevaporation with toluene. The obtained residue was dissolved in water and lyophilized. The amine acetate salt was then filtered over a basic anion-exchange resin AMBERLITE® IRA400 (OH) to remove the acetate anions. After lyophilisation, 264 mg (**44**, 3.00 mmol, 83% yield) of a yellow oil were obtained. – ^{1}H-NMR (300 MHz, MeOD): δ/ppm = 2.38 (t, J = 6.6 Hz, 2 H, *CH_2*$CONH_2$), 2.90 (t, J = 6.6 Hz, 2 H, *CH_2*NH_2). – ^{13}C-NMR (75 MHz, MeOD): δ/ppm 38.8 (*CH_2*$CONH_2$), 46.3 (*CH_2*NH_2), 177.2 (*CO*). – IR (KBr): ν (cm^{-1}) = 3357 (m), 3196 (m), 2960 (m), 1666 (s), 1411 (m), 1303 (w), 1157 (w), 1040 (w), 927 (w), 619 (w). – MS (EI, 70 eV), m/z (%): 88 (7) $[M]^+$, 72 (7) $[M-NH_2]^+$, 70 (16), 60 (39), 59 (38) $[CH_2CONH_2+H]^+$, 44 (57) $[CONH_2]^+$, 43 (100) $[COCH_2]^+$. – HRMS (EI, 70 eV): $C_3H_8N_2O$, calc. 88.0637; found: 88.0640.

***tert*-Butyl (4-aminobutyl)carbamate (81)**

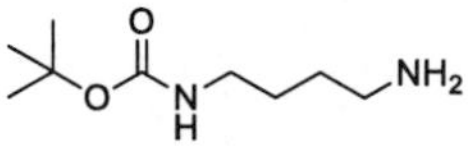

Synthesis[123] and characterization[141] of this compound have been previously described.

***tert*-Butyl (5-aminopentyl)carbamate (125)**

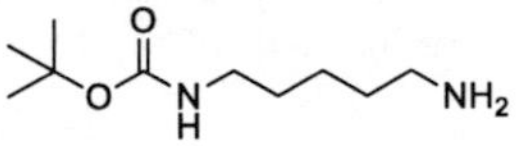

A solution of 0.58 mL Boc-anhydride (**48**, 0.59 g, 2.71 mmol, 1.00 equiv.) in 12 mL THF was added dropwise overnight under stirring at room temperature to a solution of 2.5 mL pentan-1,5-diamine (2.13 g, 20.9 mmol, 7.69 equiv.) in 25 mL THF. After completion of the reaction, the solvent was removed under reduced pressure. The residue was collected in 20 mL of water, filtered over Celite® and extracted with ethyl acetate (3 × 20 mL). The combined organic layers were washed with brine (3 × 20 mL) and dried over sodium sulfate. Removal of the solvent under reduced pressure afforded 490 mg of the product (2.42 mmol, 89% yield) as a colorless oil. – ^{1}H-NMR (300 MHz, $CDCl_3$): δ/ppm= 1.18 (bs, 2 H, NH_2), 1.31–1.53 (m, 15 H, $(CH_2)_3$ & $C(CH_3)_3$), 2.69 (t, J = 6.8 Hz, 2 H, CH_2NH_2), 3.11 (q, J = 12.8 Hz, 2 H, CH_2NHCO), 4.52 (bs, 1 H, N*H*). Analytics are consistent with those reported in the literature.[142]

2-(((4-Methoxyphenyl)diphenylmethyl)thio)ethanamine (128)

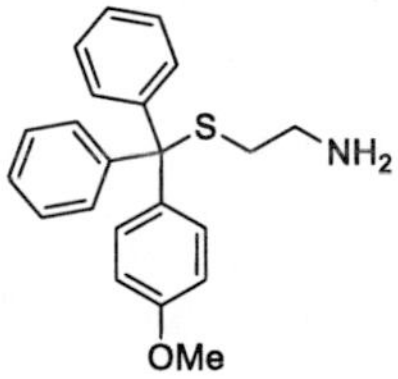

Compound **128** was synthesized following the procedure described by Valliant.[126] To a solution of 683 mg cysteamine hydrochloride (**127**, 6.01 mmol, 1.05 equiv.) in 9.00 mL trifluoroacetic acid, 1.86 g 4-methoxytriphenylchloromethane (6.01 mmol, 1.00 equiv.) were added under argon at room temperature. The reaction mixture was stirred at room temperature for 6 h and the solvent was evaporated under reduced pressure. The residue was suspended in 20 mL dichloromethane and washed with 3 M sodium hydroxide solution (3 × 10 mL). Finally the organic phase was dried over sodium sulfate and the solvent was evaporated obtaining 1.58 g of product (**128**, 4.52 mmol, 75% yield) as a colorless oil. – ^{1}H-NMR (300 MHz, $CDCl_3$): δ/ppm = 2.30 (t, J = 6.5 Hz, 2 H, CH_2S), 2.58 (t, J = 6.5 Hz, 2 H, CH_2N), 3.77 (s, 3 H, CH_3), 6.80 (t, J = 7.7 Hz, 2 H, $H_{p\text{-}Ar}$), 7,11–7,46 (m, 12 H, H_{Ar}) – ^{13}C-NMR (75 MHz, $CDCl_3$): δ/ppm = 36.41 (CH_2S), 41.19 (CH_2N), 55.37 (CH_3), 66.23 (C_qS), 113.26 (CH_{Ar}),

126.75 (CH_{Ar}), 128.00 (CH_{Ar}), 129.64 (CH_{Ar}), 130.93 (CH_{Ar}), 137.12 (C_{qAr}), 145.35 (C_{qAr}), 158.23 (C_{qAr}) – IR(KBr): ν (cm^{-1}) = 3365 (w), 3056 (m), 3030 (m), 2953 (m), 2931 (m), 2835 (m), 2043 (vw), 1957 (vw), 1900 (vw), 1816 (vw), 1774 (vw), 1606 (s), 1581 (m), 1508 (s), 1489 (s), 1462 (m), 1444 (s), 1413 (w), 1383 (w), 1297 (m), 1250 (s), 1180 (s), 1116 (w), 1081 (w), 1034 (s), 889 (m), 825 (s), 792 (m), 759 (s), 743 (s), 726 (m), 701 (s), 667 (w), 635 (w), 622 (m), 581 (s), 539 (w), 515 (w) – MS (EI, 70eV): m/z(%) = 350 (0.06) $[M+H]^+$, 290 (30) $[M–C_2H_4NH_2–CH_3]^+$, 273 (100) $[MeOTrt]^+$, 213 (54) $[M–C_2H_4NH_2–Ph–CH_3]^+$. Analytics are consistent with those reported in the literature.[126]

***(S)*-Benzyl (1-amino-1-oxopropan-2-yl)carbamate (158)**

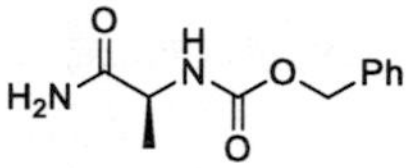

(S)-Cbz-Ala-OSu (**157**, 200 mg, 0.624 mmol, 1.00 equiv.) was suspended in 5 mL of 7 M NH_3 in MeOH and stirred overnight at room temperature. After evaporation of the solvent, the residue was resuspended in methanol and the precipitated solid was filtered off. The solvent was evaporated and the product was purified by column chromatography (cyclohexane/ethyl acetate 1:2) yielding 124 mg (0.558 mmol, 89% yield) of a white solid. – m.p. = 132–134 °C – ^{1}H-NMR (300 MHz, $CDCl_3$): δ/ppm = 1.41 (d, J = 7.0 Hz, 3 H, CH_3), 4.22-4.31 (m, 1 H, $CHCH_3$), 5.12 (s, 2 H, CH_2Ph), 5.27 (bs, 1 H, N*H*), 5.36 (bs, 1 H, N*H*), 5.96 (bs, 1 H, N*H*), 7.32-7.38 (m, 5 H, CH_{Ar}). – ^{13}C-NMR (100 MHz, $CDCl_3$): δ/ppm = 18.4 (CH_3), 50.1 (CH_2), 67.1 (*C*H), 128.1 (C_{Ar}), 128.3 (C_{Ar}), 128.6 (C_{Ar}), 136.0 (C_q, C_{Ar}), 156.0 (C_q, CO_2NH), 174.5 (C_q, $CONH_2$). – IR (ATR): ν (cm^{-1}) = 3373 (w), 3322 (w), 3191 (w), 2986 (vw), 2938 (vw), 1684 (m), 1644 (m), 1523 (m), 1461 (w), 1424 (w), 1377 (w), 1314 (m), 1288 (w), 1252 (m), 1231 (m), 1108 (w), 1064 (m), 1038 (w), 967 (vw), 928 (vw), 841 (vw), 801 (w), 780 (w), 747 (m), 725 (w), 695 (m), 630 (m), 580 (m), 554 (w), 493 (w), 433 (w). – MS (EI, 70 eV), m/z (%): 222 (13) $[M]^+$, 178 (31) $[M–CONH_2]^+$, 91 (100) $[PhCH_2]^+$, 44 (58) $[CONH_2]^+$ – HRMS (EI, 70 eV): $C_{11}H_{14}N_2O_3$, calc. 222.1004; found 222.1004. $[\alpha]^{25}_D$ = -3.5° (c = 2 in MeOH).

(*S*)-Alaninamide (155)

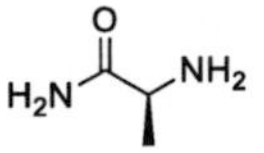

1.25 g of amide **158** (5.63 mmol, 1.00 equiv.) were dissolved in 30 mL methanol and 10 mass% of the catalyst (5% Pd/C) was added. The obtained suspension was treated with H_2 at room temperature overnight. After completion of the reaction, the catalyst was removed by filtration over Celite® and the solvent was evaporated yielding 480 mg (5.45 mmol, 97%) of a colorless oil that solidified with time. – ^{1}H-NMR (300 MHz, MeOD): δ/ppm = 1.28 (d, J = 6.9 Hz, 3 H, CH_3), 3.41 (q, J = 6.9 Hz, 1 H, C*H*CH_3). ^{13}C-NMR (100 MHz, MeOD): δ/ppm = 21.6 (CH_3), 51.3 (*C*H), 181.3 (C_q, *C*ONH$_2$). – IR (KBr): ν (cm^{-1}) = 3358 (m), 3203 (m), 2978 (m), 2936 (w), 1668 (m), 1458 (w), 1414 (w), 1371 (w), 1242 (vw), 1134 (vw), 1073 (vw), 960 (vw), 910 (vw), 638 (w). – MS (EI, 70 eV), m/z (%): 88 (0.8) $[M]^+$, 73 (1.2) $[M-CH_3]^+$, 55 (1), 44 (100) $[CONH_2]^+$. – HRMS (EI, 70 eV): $C_3H_8N_2O$, calc. 88.0637; found 88.0638. $[\alpha]^{21}_D$ = +6.6° (c = 1 in MeOH).

Peptoid syntheses

H-(N1ph)$_2$-NH-SP (68)

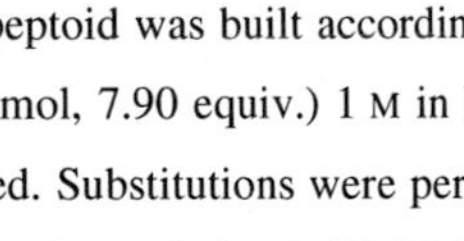

905 mg of low-loading Rink Amide resin (0.308 mmol, 1.00 equiv.) were swollen and deprotected following **GP1**. The peptoid was built according to **GP3**. For the acylation steps 338 mg bromoacetic acid (2.43 mmol, 7.90 equiv.) 1 M in DMF and 0.38 mL DIC (306 mg, 2.43 mmol, 7.90 equiv.) were used. Substitutions were performed with 336 μL benzylamine (330 mg, 3.08 mmol, 10.0 equiv.) as 1 M solution in DMF. Finally, the resin was washed with dichloromethane (3 × 5 mL) and dried overnight.

H-N2mo-N2ad-(N1ph)$_2$-NH$_2$ (70)

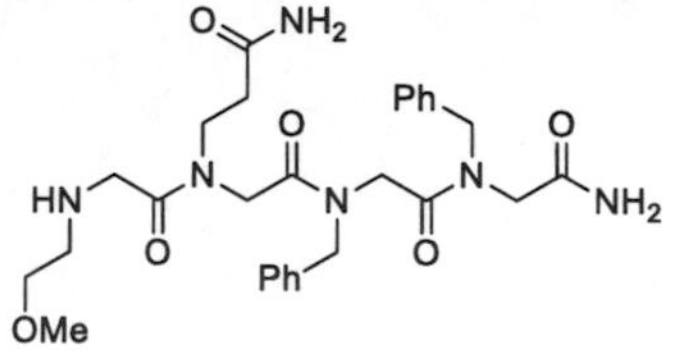

200 mg (66.0 μmol, 1.00 equiv.) of the resin-bound peptoid **68** were swollen in 2 mL DMF for 2 h. The peptoid was built as described in **GP3**. For the acylation steps 83 mg bromoacetic acid (0.594 mmol, 9.0 equiv.) 1 M in DMF and 72 μL DIC (58 mg, 0.462 mmol, 7.00 equiv.) were used. Substitutions were performed with 48 mg amine **44** (0.548 mmol, 8.30 equiv.) 1 M in NMP and 48 μL methoxyethylamine (41 mg, 0.548 mmol, 8.30 equiv.) 1 M in DMF. The peptoid was cleaved from the resin with 2 mL 95% TFA in dichloromethane according to **GP12** (2 × 2 h). HPLC purification afforded 28.9 mg peptoid **70** (52.1 μmol, 79% yield). In addition, 2.3 mg dimer **73** (2.9 μmol, 8.8% yield) could be isolated.

Peptoid **70:** MALDI-TOF (matrix: DHB, 0.1% TFA) m/z: 555 $[M+H]^+$.– Analytical HPLC (5–95% acetonitrile in water + 0.1% TFA over 20 min): t_R =10.6 min, 92% purity.

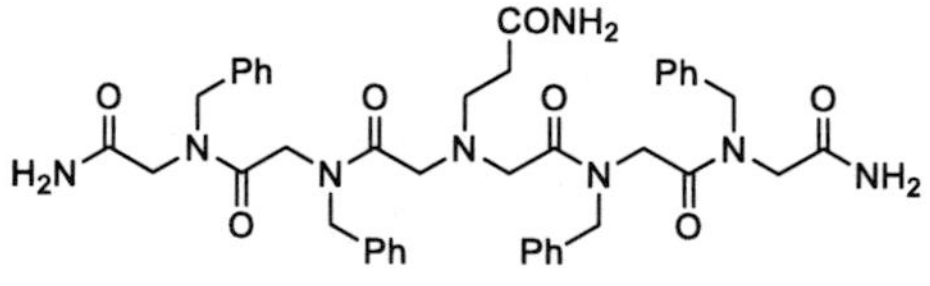

73

Dimer **73**: MALDI-TOF (matrix: DHB, 0.1% TFA) m/z : 791 $[M+H]^+$.– Analytical HPLC (5–95% acetonitrile in water + 0.1% TFA over 20 min): t_R = 13.4 min, 78% purity.

H-N2mo-N3ad-(N1ph)$_2$-NH$_2$ (71)

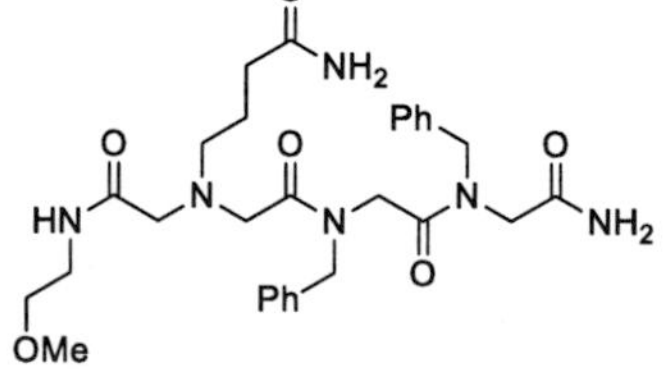

200 mg (66.0 μmol, 1.00 equiv.) of the resin bound peptoid **68** were swollen in 2 mL DMF for 2 h. The peptoid was built as described in **GP3**. For the acylation steps 83 mg bromoacetic

acid (0.594 mmol, 9.0 equiv.) 1 M in DMF and 72 µL DIC (58 mg, 0.462 mmol, 7.00 equiv.) were used. Substitutions were performed with 56 mg amine **45** (0.548 mmol, 8.30 equiv.) 1 M in NMP and 48 µL methoxyethylamine (41 mg, 0.548 mmol, 8.30 equiv.) 1 M in DMF. The peptoid was cleaved from the resin with 2 mL 95% TFA in dichloromethane according to **GP12** (2 × 2 h). HPLC purification afforded 30.0 mg peptoid **71** (52.8 µmol, 80% yield). In addition 2.2 mg dimer **74** (2.7 µmol, 8.3% yield) could be isolated.

Peptoid **71**: MALDI-TOF (matrix: DHB, 0.1% TFA) m/z: 569 $[M+H]^+$.– Analytical HPLC (5–95% acetonitrile in water + 0.1% TFA over 20 min): t_R =10.8 min, 95% purity.

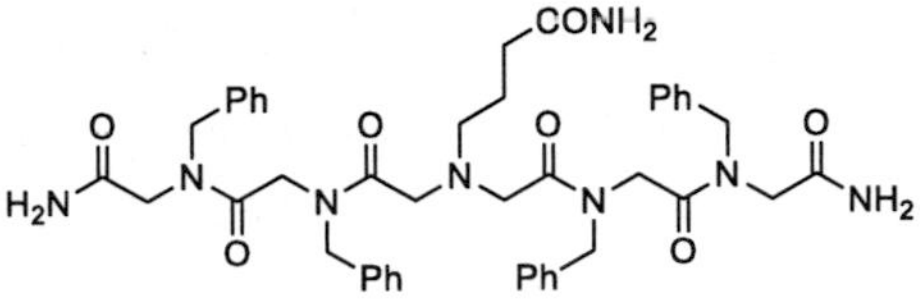

74

Dimer **74**: MALDI-TOF (matrix: DHB, 0.1% TFA) m/z : 805 $[M+H]^+$.– Analytical HPLC (5–95% acetonitrile in water + 0.1% TFA over 20 min): t_R=13.4 min, 85% purity.

Ac-N2ad-N4gn-N2in-N2im-N3cx-NH$_2$ (80)

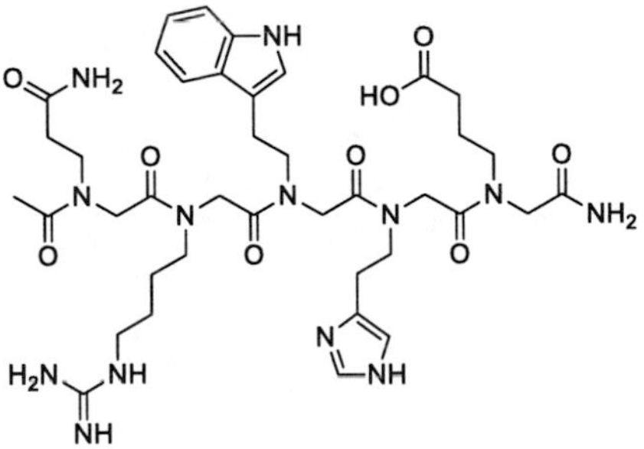

100 mg low-loading Rink-amide resin (0.039 mmol, 1.00 equiv.) were swollen and deprotected following **GP1**. Peptoid residues N3cx^{tBu} and N2im were introduced following **GP6**. For the acylation steps 49 mg bromoacetic acid (0.351 mmol, 9.00 equiv.) 0.6 M in DMF and 42 µL DIC (34 mg, 0.273 mmol, 7.00 equiv.) were used. Each reaction was shaken for 20 min. Substitutions were performed with amine **43** (47 mg, 0.295 mmol, 7.56 equiv.) and histamine (**41**, 36 mg, 0.324 mmol, 8.30 equiv.) dissolved in 0.5 mL DMF and shaken for 1 h 15 min and 2 h 15 min, respectively. The following side-chains were coupled to the resin

according to **GP7**. For the acylation steps 33 mg chloracetic acid (0.351 mmol, 9.00 equiv.) 0.6 M in DMF and 42 µL DIC (34 mg, 0.273 mmol, 7.00 equiv.) were used. The reactions were shaken for 2–3 h. Substitutions were performed with tryptamine (**39**, 53 mg, 0.324 mmol, 8.30 equiv.) and unprotected 1,4-diaminobutane (**51**, 29 mg, 0.324 mmol, 8.30 equiv.), each dissolved in 0.6 mL DMF and shaken for 3 h 30 min and 1 h 30 min, respectively. After coupling with 1,4-diaminobutane this residue was guanidinated on solid phase overnight according to **GP9** using 60 mg *N,N'*-di-Boc-1*H*-pyrazole-1-carboxamidine (**50**, 0.195 mmol, 5.00 equiv.) and 0.05 mL DIPEA (38 mg, 0.293 mmol, 7.50 equiv.) dissolved in 1 mL DMF. Submonomer N2ad was also incorporated as described in **GP7** with the same acylation conditions used for the previous couplings. Amine **44** (44 mg, 0.499 mmol, 12.8 equiv.) was employed for the substitution step and the reaction was shaken overnight. Finally the peptoid was capped overnight following **GP11** with 37 µL acetic anhydride (40 mg, 0.390 mmol, 10.0 equiv.) and 99 µL DIPEA (76 mg, 0.585 mmol, 15.0 equiv.) in 1 mL DMF. Then the peptoid was cleaved from the resin with 1 mL 95% TFA solution in dichloromethane, as described in **GP12** (2 × 3 h). After HPLC purification 2.0 mg peptoid **80** (2.35 µmol, 6% yield over 14 steps) were isolated.

MALDI-TOF (matrix: DHB, 0.1% TFA) m/z: 852 $[M+H]^+$.– Analytical HPLC (5–95% acetonitrile in water + 0.1% TFA over 20 min): t_R = 8.7–9.0 min, 38–58% purity.

H-N2ad-N4gn-N2in-N2im-N3cx-NH_2 (38)

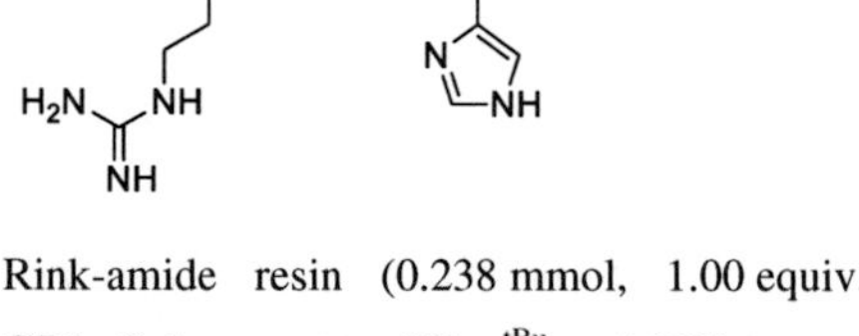

700 mg low-loading Rink-amide resin (0.238 mmol, 1.00 equiv.) were swollen and deprotected following **GP1**. Submonomers $N3cx^{tBu}$ and N2im were coupled according to **GP3**. The acylations were performed with 261 mg bromoacetic acid (1.88 mmol, 7.90 equiv.) 1 M in DMF and 0.48 mL DIC (388 mg, 3.08 mmol, 12.9 equiv.). For the substitution reactions, amine **43** (299 mg, 1.88 mmol, 7.90 equiv.) and histamine (**41**, 209 mg, 1.88 mmol,

7.90 equiv.), both as 1 M solution in DMF, were employed. Then, submonomers N2in, N4am^{Boc}, and N2ad were incorporated as described in **GP4**. The acylations were performed with 202 mg chloracetic acid (2.14 mmol, 9.00 equiv.) 1 M in DMF and 0.33 mL DIC (270 mg, 2.14 mmol, 9.00 equiv.). For the substitution reactions the following amines were used: tryptamine (**39**, 343 mg, 2.14 mmol, 9.00 equiv.) 1 M in DMF, Boc-protected diaminobutane **81** (403 mg, 2.14 mmol, 9.00 equiv.) 1 M in DMF and amine **44** (189 mg, 2.14 mmol, 9.00 equiv.) 1 M in NMP. Finally the peptoid was cleaved from the resin with 4 mL 95% TFA solution in dichloromethane, as described in **GP12** (2 × 2 h). Guanidination of the amine side-chain was carried out in solution according to **GP10** with 105 mg 1*H*-pyrazole-1-carboxamidine monohydrochloride (0.714 mmol, 3.00 equiv.) and 0.37 mL DIPEA (277 mg, 2.14 mmol, 9.00 equiv.) in 2.1 mL DMA. HPLC purification afforded 6.7 mg peptoid **38** (8.3 µmol, 3.5% yield over 13 steps).

MALDI-TOF (matrix: DHB, 0.1% TFA) m/z: 810 $[M+H]^+$.– Analytical HPLC (5–95% acetonitrile in water + 0.1% TFA over 20 min): t_R = 7.8 min, 95% purity.

Ac-N3ad-Gly-N2in-N3ad-Gly-NH₂ (Ac-36)

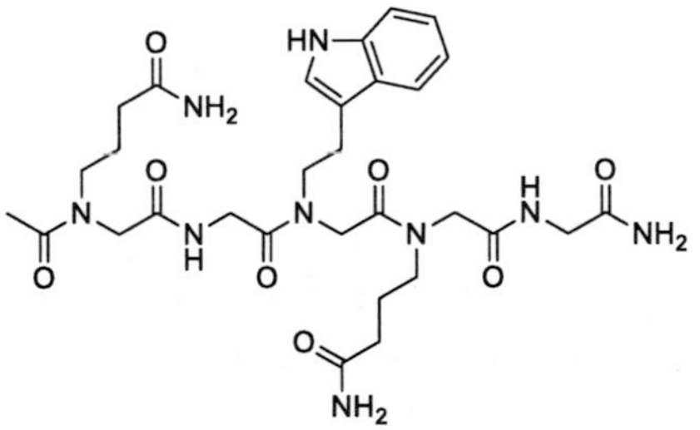

123 mg low-loading Rink-amide resin (41.8 µmol, 1.00 equiv.) were swollen and deprotected following **GP1**. Fmoc-glycine (37 mg, 0.125 mmol, 3.00 equiv.) was coupled to the resin as described in **GP2** with 19 mg HOBt·H_2O (0.125 mmol, 3.00 equiv.) and 20 µL DIC (16 mg, 0.125 mmol, 3.00 equiv.) in 0.6 mL DMF. After deprotection, submonomers N3ad and N3in were introduced according to **GP3**. Acylations were performed with 52 mg bromoacetic acid (0.376 mmol, 9.00 equiv.) 1 M in DMF and 46 µL DIC (37 mg, 0.293 mmol, 7.00 equiv.). For the substitution reactions, amine **45** (35 mg, 0.347 mmol, 8.3 equiv.) 1 M in DMF, and tryptamine (**39**, 67 mg, 0.418 mmol, 10.00 equiv.) 1 M in DMF, were employed. Then Fmoc-glycine was coupled again as described above. After deprotection submonomer N3ad was incorporated as described in **GP4**. The acylation was performed with 36 mg chloracetic acid

(0.376 mmol, 9.00 equiv.) 1 M in DMF and 0.46 mL DIC (37 mg, 0.293 mmol, 9.00 equiv.). For the substitution amine **45** (43 mg, 0.418 mmol, 10.0 equiv.) 1 M in NMP was used. Finally the peptoid was capped overnight following **GP11** with 39 µL acetic anhydride (43 mg, 0.418 mmol, 10.0 equiv.) and 107 µL DIPEA (81 mg, 0.627 mmol, 15.0 equiv.) in 0.5 mL DMF. The cleavage from the resin was performed as described in **GP12** with 2 mL 95% TFA solution in dichloromethane for 2.5 and 1.5 h. HPLC purification afforded 1.3 mg peptoid **Ac-36** (2.0 µmol, 4.7% yield over 13 steps).

MALDI-TOF (matrix: DHB/CHCA, 0.1% TFA) m/z: 680 $[M+Na]^+$, 696 $[M+K]^+$.– Analytical HPLC (5–95% acetonitrile in water + 0.1% TFA over 20 min): t_R = 9.0 min, 80% purity.

Ac-N3cx-N2im-N2in-N4gn-N2ad-NH_2 (87)

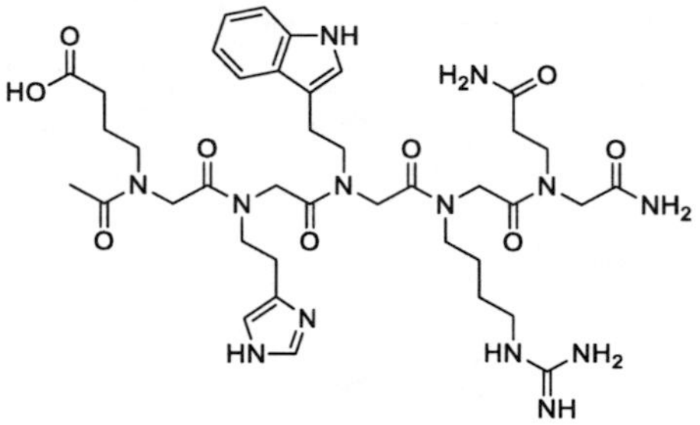

102 mg low-loading Rink-amide resin (0.035 mmol, 1.00 equiv.) were swollen and deprotected following **GP1**. Submonomers N2ad, N4am^{Boc} and N2in were introduced following **GP3**. For the acylation steps 24 mg bromoacetic acid (0.274 mmol, 7.90 equiv.) 0.9 M in DMF and 70 µL DIC (57 mg, 0.274 mmol, 7.90 equiv.) were used. Substitutions were performed with amine **44** (24 mg, 0.274 mmol, 7.90 equiv.) 0.9 M in DMF, Boc-protected diaminobutane **81** (52 mg, 0.274 mmol, 7.90 equiv.) 0.9 M in DMF, and tryptamine (**39**, 44 mg, 0.274 mmol, 7.90 equiv.) 0.7 M in DMF. Then side-chains N2im and N3cx^{tBu} were coupled to the resin according to **GP4**. For the acylation steps 29 mg chloracetic acid (0.312 mmol, 9.00 equiv.) 1 M in DMF and 50 µL DIC (39 mg, 0.312 mmol, 9.00 equiv.) were used. Substitutions were performed with histamine (**41**, 35 mg, 0.312 mmol, 9.00 equiv.) and amine **43** (50 mg, 0.312 mmol, 9.00 equiv.), each as 1 M solution in DMF. Finally the peptoid was capped overnight following **GP11** with 33 µL acetic anhydride (35 mg, 0.347 mmol, 10.0 equiv.) and 88 µL DIPEA (67 mg, 0.520 mmol, 15.0 equiv.) in 0.5 mL DMF. Cleavage from the resin was performed as described in **GP12** with 1 mL 95% TFA

solution in dichloromethane for 2 and 4 h. After lyophilization the amine side-chain was guanidinated according to **GP10** with 15 mg 1*H*-pyrazole-1-carboxamidine hydrochloride (**84**, 0.105 mmol, 3.00 equiv.) and 54 µL DIPEA (41 mg, 0.315 mmol, 9.00 equiv.) in 2 mL DMA. The peptoid was purified by HPLC affording 2.1 mg of peptoid **87** (2.5 µmol, 7% yield over 14 steps).

MALDI-TOF (matrix: DHB/CHCA 1:1, 0.1% TFA) m/z: 852 $[M+H]^+$, 874 $[M+Na]^+$, 908 $[M+K]^+$ – Analytical HPLC (5–95% acetonitrile in water + 0.1% TFA over 20 min): t_R = 8.8 min, >99% purity.

H-Gly-N4am-N2in-N2im-Gly-NH₂ (88)

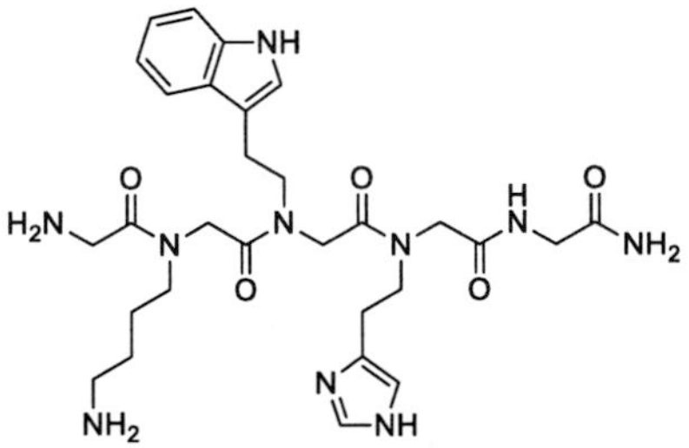

152 mg low-loading Rink-amide resin (0.059 mmol, 1.00 equiv.) were swollen and deprotected following **GP1**. Fmoc-glycine (53 mg, 0.178 mmol, 3.00 equiv.) was coupled as described in **GP5** (2 × 2 h). For each coupling 27 mg HOBt·H_2O (0.178 mmol, 3.00 equiv.) and 28 µL DIC (22 mg, 0.178 mmol, 3.00 equiv.) 0.1 M in DMF were used. Submonomer N2im was introduced according to **GP6**. For the acylation step 74 mg bromoacetic acid (0.534 mmol, 9.00 equiv.) 0.5 M in DMF and 65 µL DIC (52 mg, 0.415 mmol, 7.00 equiv.) were used and the reaction was shaken for 30 min. The substitution step was performed with histamine (**41**, 55 mg, 0.492 mmol, 8.30 equiv.) 1 M in DMF shaking overnight. Residues N2in and N4am^{Boc} were coupled to the resin according to **GP7**. The acylations were performed with 50 mg chloracetic acid (0.534 mmol, 9.00 equiv.) 0.5 M in DMF and 65 µL DIC (52 mg, 0.415 mmol, 7.00 equiv.). The reactions were shaken for 1 h 40 min and 3 h 25 min, respectively. For the substitution reactions tryptamine (**39**, 80 mg, 0.492 mmol, 8.30 equiv.) and Boc-diaminobutane **81** (94 mg, 0.492 mmol, 8.30 equiv.) each as 1 M solution in DMF were employed. The reactions were shaken for 3 h and 15 h, respectively. At the *N*-terminus Fmoc-glycine was coupled again as described above. After Fmoc-deprotection, the peptoid was cleaved from the resin with 1 mL 95% TFA solution in

dichloromethane, as described in **GP12** (2 × 2 h). HPLC purification afforded 5.2 mg of peptoid **88** (8.68 μmol, 15% yield over 12 steps).

MALDI-TOF (matrix: DHB, 0.1% TFA) m/z: 611 $[M+H]^+$.– Analytical HPLC (5–95% acetonitrile in water + 0.1% TFA over 20 min): t_R = 7.3 min, 97% purity.

H-Nspm-N4gn-N2in-N2im-Nspm-NH_2 (89)

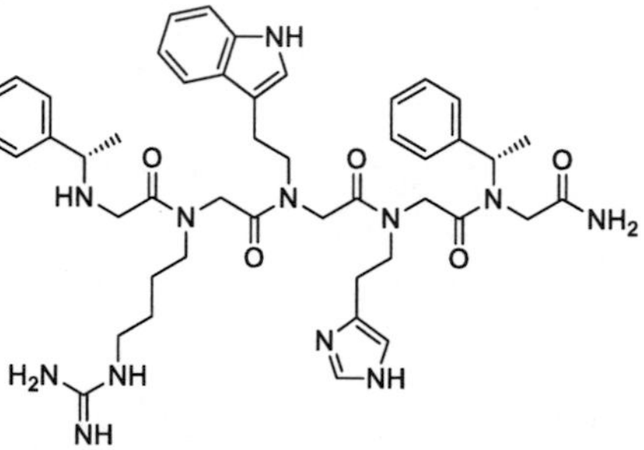

152 mg low-loading Rink-amide resin (0.052 mmol, 1.00 equiv.) were swollen and deprotected following **GP1**. Residues Nspm and N2im were introduced following **GP3**. For the acylation steps 57 mg bromoacetic acid (0.408 mmol, 7.90 equiv.) 1 M in DMF and 64 μL DIC (51 mg, 0.408 mmol, 7.90 equiv.) were used. Substitutions were performed with (*S*)-1-phenylethanamine (56 mg, 0.465 mmol, 9.00 equiv.) and histamine (**41**, 52 mg, 0.465 mmol, 9.00 equiv.) as 1.2 M solutions in DMF. Then, side-chains N2in, $N4am^{Boc}$ and again Nspm were coupled to the resin according to **GP4**. For the acylation steps 44 mg chloracetic acid (0.465 mmol, 9.00 equiv.) 1.2 M in DMF and 64 μL DIC (51 mg, 0.408 mmol, 7.90 equiv.) were used. Substitutions were performed with tryptamine (**39**, 76 mg, 0.465 mmol, 9.00 equiv.), Boc-protected diaminobutane **81** (89 mg, 0.465 mmol, 9.00 equiv.), and (*S*)-1-phenylethanamine (56 mg, 0.465 mmol, 9.00 equiv.), each as 1.2 M solution in DMF. Finally the peptoid was cleaved from the resin with 1 mL 95% TFA solution in dichloromethane, as described in **GP12** (2 × 2 h) and lyophilized. Guanidination of the amine side-chain was carried out in solution according to **GP10** with 23 mg 1*H*-pyrazole-1-carboxamidine hydrochloride (0.156 mmol, 3.00 equiv.) and 0.08 mL DIPEA (60 mg, 0.468 mmol, 9.00 equiv.) in 1 mL DMA. HPLC purification afforded 0.8 mg peptoid **89** (0.930 μmol, 1.8% yield over 13 steps).

MALDI-TOF (matrix: DHB/CHCA, 0.1% TFA) m/z: 852 $[M+H]^+$. – Analytical HPLC (5–95% acetonitrile in water + 0.1% TFA over 20 min): t_R = 10.8 min, 99% purity.

Ac-N2ad-Arg-Trp-His-N3cx-NH$_2$ (90)

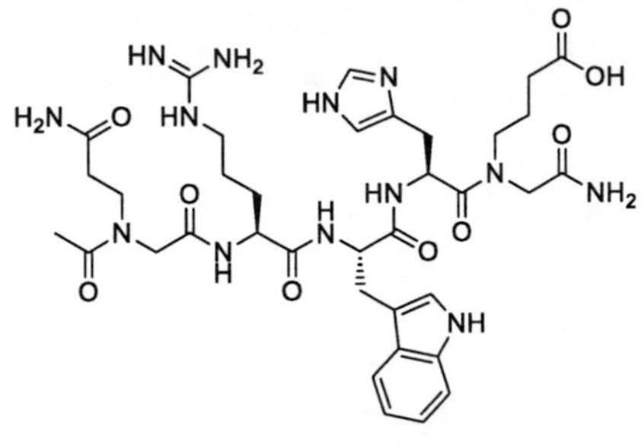

202 mg low-loading Rink-amide resin (0.069 mmol, 1.00 equiv.) were swollen and deprotected following **GP1**. First the peptoid residue N3cx^{tBu} was introduced following **GP3**. For the acylation step 75 mg bromoacetic acid (0.543 mmol, 7.90 equiv.) 1 M in DMF and 85 µL DIC (68 mg, 0.543, 7.90 equiv.) were used. Substitution was performed with amine **43** (86 mg, 0.543 mmol, 7.90 equiv.) as 1 M solution in DMF. Then, the central amino acids were coupled to the resin according to **GP2**. They were incorporated as protected monomers Fmoc-His(Trt)-OH (128 mg, 0.206 mmol, 3.00 equiv.), Fmoc-Trp(Boc)-OH (108 mg, 0.206 mmol, 3.00 equiv.) and Fmoc-Arg(Pbf)-OH (134 mg, 0.206 mmol, 3.00 equiv.). For each coupling 32 mg HOBt·H_2O (0.206 mmol, 3.00 equiv.), 32 µL DIC (26 mg, 0.206 mmol, 3.00 equiv.) and 1 mL DMF were used. After Fmoc deprotection of the *N*-terminus, the peptoid residue N2ad was incorporated following **GP3**. The acylation was performed as described above. For the substitution step amine **44** (48 mg, 0.543 mmol, 7.90 equiv.) as 1M solution in DMF was employed. Finally the peptoid was capped overnight following **GP11** with 65 µL acetic anhydride (70 mg, 0.69 mmol, 10.0 equiv.) and 0.18 mL DIPEA (133 mg, 1.03 mmol, 15.0 equiv.) in 1 mL DMF. Cleavage from resin was performed with 2 mL 95% TFA solution in dichloromethane, as described in **GP12** (2 × 2.5 h). After HPLC purification 1.4 mg peptoid **90** (1.7 µmol, 2.5% yield over 14 steps) were isolated.

MALDI-TOF (matrix: DHB/CHCA 1:1, 0.1% TFA) m/z: 810 $[M+H]^+$, 832 $[M+Na]^+$. – Analytical HPLC (5–95% acetonitrile in water + 0.1% TFA over 20 min): t_R = 7.9 min, 95% purity.

H-N1ph-(N4am^{Boc}-N2in-N2im-)$_2$-N1ph-OH (100)

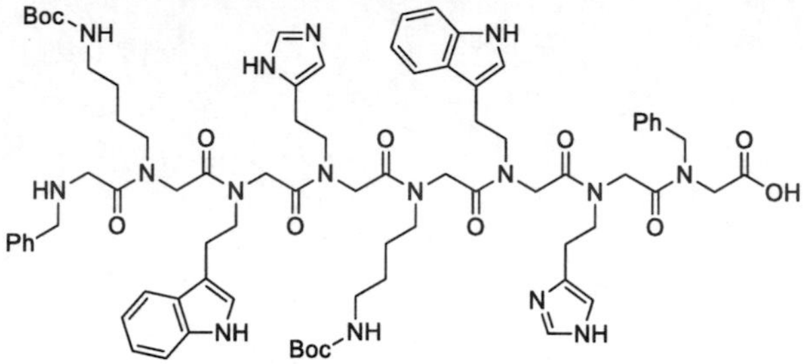

101 mg 2-chlorotrityl chloride resin (0.208 mmol, 1.00 equiv.) were treated as described in **GP8**. For the first reaction 38 mg bromoacetic acid (0.270 mmol, 1.30 equiv.) and 0.18 mL DIPEA (140 mg, 1.08 mmol, 5.20 equiv.) dissolved in 0.5 mL dichloromethane were used. The mixture was shaken for 1.5 h. The next steps were performed according to **GP6**: the first substitution was performed with benzylamine (223 mg, 2.08 mmol, 10.0 equiv.) 1 M in DMF for 2 h 45 min, followed by acylation with 260 mg bromoacetic acid (1.87 mmol, 9.00 equiv.) 1 M in DMF and 0.23 mL DIC (184 mg, 1.46 mmol, 7.00 equiv.) for 2 h and a second substitution with histamine (**41**, 231 mg, 2.08 mmol, 10.0 equiv.) 1.2 M in DMF overnight. The rest of the couplings were carried out as described in **GP7**. For the acylations 177 mg chloracetic acid (1.87 mmol, 9.00 equiv.) 1 M in DMF and 0.26 mL DIC (207 mg, 1.64 mmol, 7.90 equiv.) were used. The reactions were shaken for 2–4 h. Substitutions were performed in the following order: tryptamine (**39**, 306 mg, 1.87 mmol, 9.00 equiv.) 1 M in DMF for 2 h, Boc-protected diaminobutane **81** (285 mg, 1.52 mmol, 7.30 equiv) 1 M in DMF overnight, histamine (**41**, 208 mg, 1.87 mmol, 9.00 equiv.) 1 M in DMF overnight, tryptamine (**39**, 306 mg, 1.87 mmol, 9.00 equiv.) 1 M in DMF overnight, Boc-protected diaminobutane **81** (352 mg, 1.87 mmol, 9.00 equiv.) 1 M in DMF for 2 h, and benzylamine (223 mg, 2.08 mmol, 10.0 equiv.) 1.1 M in DMF for 2.5 h. The cleavage from the resin was performed according to **GP13** with 1 mL 20% HFIP in dichloromethane for 1 and 2 h, obtaining 104 mg (71 µmol, 34% yield over 17 steps) of crude peptoid **100**, which was cyclized without further purification.

MALDI-TOF (matrix: DHB/CHCA, 0.1% TFA) m/z: 1473 $[M+H]^+$, 1494 $[M+Na]^+$.

c-(N1ph)$_2$-(N4am-N2in-N2im)$_2$ (102b)

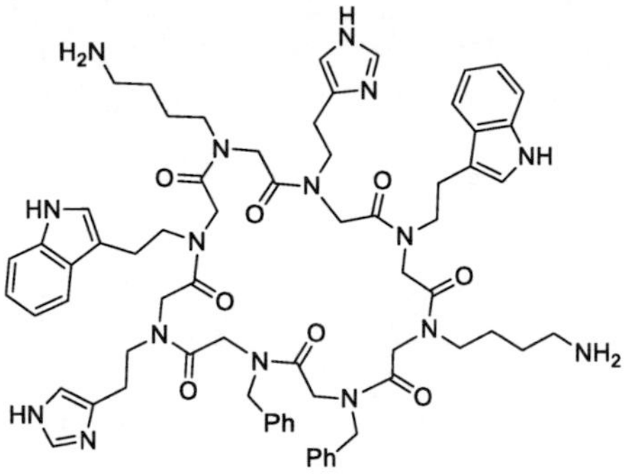

The crude linear peptoid **101** (104 mg, 71 µmol, 1.00 equiv.) was cyclized following **GP14** with 109 mg PyBOP (0.210 mmol, 3.00 equiv.) and 71 µL DIPEA (54 mg, 0.420 mmol, 6.00 equiv.) in 35 mL dry DMF for 1 h. Then the Boc-protecting groups were removed by stirring the peptoid in 10 mL TFA/dichloromethane (1:1) solution for 3 h. The crude was pre-purified by HPLC affording 19 mg peptoid **102** (15 µmol, 7.2% yield over 19 steps).

MALDI-TOF (matrix: DHB/CHCA, 0.1% TFA) m/z: 1254 $[M+H]^+$.

c-(N1ph)$_2$-(N4gn-N2in-N2im)$_2$ (91)

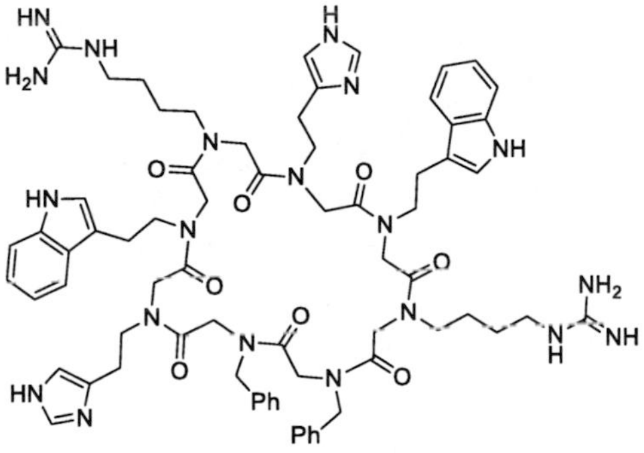

The purified peptoid **102b** (19 mg, 15 µmol, 1.00 equiv.) was guanidinated according to **GP10** with 13 mg 1*H*-pyrazole-1-carboxamidine hydrochloride (0.091 mmol, 6.00 equiv.) and 47 µL DIPEA (35 mg, 0.273 mmol, 18.00 equiv.) in 1.5 mL DMA. The peptoid was purified *via* HPLC affording 5.3 mg peptoid **91** (3.96 µmol, 26% guanidination yield, 1.8% yield over 20 steps).

MALDI-TOF (matrix: DHB, 0.1% TFA) m/z: 1337 $[M+H]^+$, 1360 $[M+Na]^+$.– Analytical HPLC (5–95% acetonitrile in water + 0.1% TFA over 20 min): t_R = 11.0 min, 98% purity.

H-(N2in-N2im-N4am^{Boc})$_2$-OH (101)

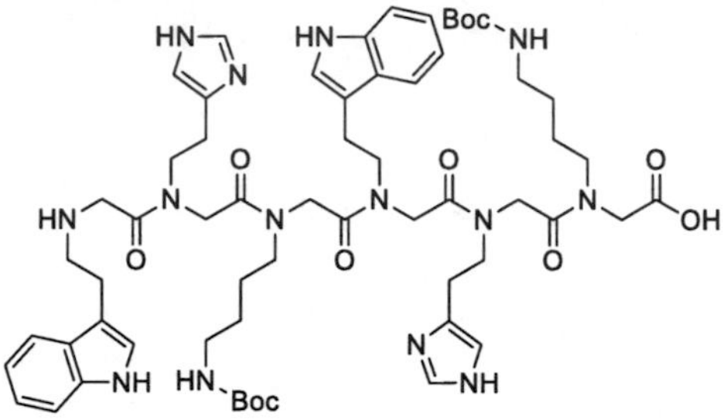

121 mg 2-chlorotrityl chloride resin (0.249 mmol, 1.00 equiv.) were treated as described in **GP8**. For the first reaction 45 mg bromoacetic acid (0.324 mmol, 1.30 equiv.) and 0.22 mL DIPEA (168 mg, 1.30 mmol, 5.20 equiv.) dissolved in 0.6 mL dichloromethane were used. The mixture was shaken for 1 h. The next steps were performed according to **GP6**: the substitution step was performed with Boc-protected diaminobutane **81** (389 mg, 2.07 mmol, 8.30 equiv.) 1 M in DMF for 2 h, the acylation with 312 mg bromoacetic acid (2.24 mmol, 9.00 equiv.) 1 M in DMF and 0.27 mL DIC (220 mg, 1.75 mmol, 7.00 equiv.) for 1 h, and the second substitution with histamine (**41**, 277 mg, 2.49 mmol, 10.0 equiv.) 1.2 M in DMF overnight. The rest of the couplings were carried out as described in **GP7**. For the acylations 212 mg chloracetic acid (2.243 mmol, 9.00 equiv.) 1 M in DMF and 0.3 mL DIC (248 mg, 1.969 mmol, 7.90 equiv.) were used. The reactions were shaken for 2.5 h. Substitutions were performed with tryptamine (**39**, 367 mg, 2.243 mmol, 9.00 equiv.) 1 M in DMF for 2 h 15 min, Boc-protected diaminobutane **81** (422 mg, 2.243 mmol, 9.00 equiv.) 1 m in DMF overnight, histamine (**41**, 277 mg, 2.493 mmol, 10.0 equiv.) 1.2 m in DMF overnight, and tryptamine (**39**, 367 mg, 2.243 mmol, 9.00 equiv.) again 1 M in DMF overnight. Finally, the peptoid was cleaved from the resin according to **GP13** with 1 mL 20% HFIP in dichloromethane (2 × 2 h). The compound was cyclized without previous purification.

MALDI-TOF (matrix: DHB/CHCA 1:1, 0.1% TFA) m/z: 1177 $[M+H]^+$.

c-(N2in-N2im-N4am^{Boc})$_2$ (103)

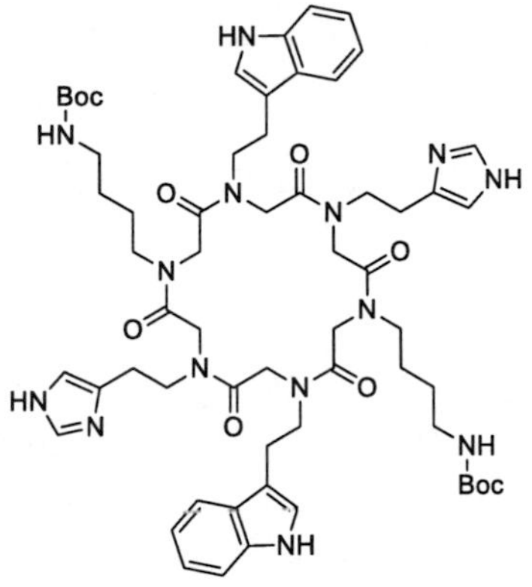

Crude linear peptoid **101** (0.249 mmol, 1.00 equiv.) was cyclized following **GP14** with 389 mg PyBOP (0.747 mmol, 3.00 equiv.) and 0.25 mL DIPEA (193 mg, 1.494 mmol, 6.00 equiv.) in 100 mL dry DMF for 1 h.

Half of the obtained crude was used in the next reaction without purification. The other half was purified by HPLC affording 10.8 mg peptoid **103** (9.3 µmol, 7.4% yield over 14 steps, calculated on half assay).

MALDI-TOF (matrix: DHB/CHCA, 0.1% TFA) m/z: 1160 $[M+H]^+$.– Analytical HPLC (5–95% acetonitrile in water + 0.1% TFA over 20 min): t_R = 13.1 min, 56–78% purity.

c-(N2in-N2im-N4gn)$_2$ (92)

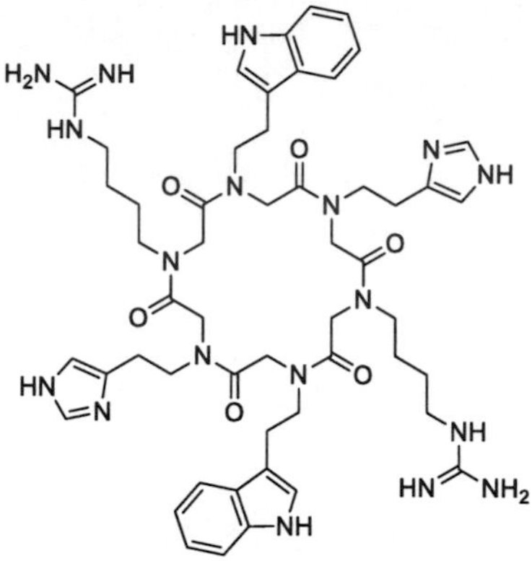

Purified peptoid **103** (10.8 mg, 9.3 µmol, 1.00 equiv.) was dissolved in 4 mL TFA/dichloromethane (1:1) solution and stirred for 4 h. After removal of the solvent the peptoid was guanidinated according to **GP10** with 8 mg 1*H*-pyrazole-1-carboxamidine hydrochloride (0.056 mmol, 6.00 equiv.) and 29 µL DIPEA (22 mg, 0.167 mmol, 18.00 equiv.) in 2 mL DMA. HPLC purification afforded 0.6 mg peptoid **92** (0.58 µmol, 6.2% yield from crude starting material; 0.5% yield over 16 steps).

MALDI-TOF (matrix: DHB, 0.1% TFA) m/z: 1337 $[M+H]^+$, 1360 $[M+Na]^+$.– Analytical HPLC (5–95% acetonitrile in water + 0.1% TFA over 20 min): t_R = 9.1 min, 98% purity.

H-(Sar-N2mo)$_3$-NH$_2$ (107)

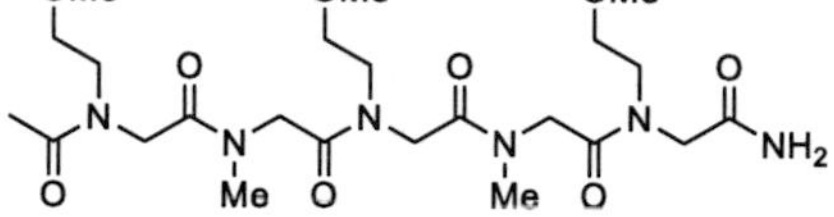

206 mg low-loading Rink-amide resin (59.8 µmol, 1.00 equiv.) were swollen and deprotected as described in **GP1**. N2mo side-chains were introduced according to **GP3**. Acylations were performed with 66 mg bromoacetic acid (0.472 mmol, 7.90 equiv.) 1 M in DMF and 74 µL DIC (60 mg, 0.472 mmol, 7.90 equiv.). For the substitution reactions 41 µL methoxyethylamine (35 mg, 0.472 mmol, 7.90 equiv.) 1 M in DMF were employed. Fmoc-sarcosine (56 mg, 0.179 mmol, 3.00 equiv.) was coupled to the resin as described in **GP2** with 27 mg HOBt·H_2O (0.179 mmol, 3.00 equiv.) and 28 µL DIC (23 mg, 0.179 mmol, 3.00 equiv.) in 0.9 mL DMF. Methoxyethylamine and sarcosine couplings were alternated

until the desired peptoid length was reached. Finally the peptoid was capped overnight following **GP11** with 56 µL acetic anhydride (61 mg, 0.598 mmol, 10.0 equiv.) and 153 µL DIPEA (116 mg, 0.897 mmol, 15.0 equiv.) in 0.5 mL DMF. The cleavage from the resin was performed according to **GP12** with 2 mL 95% TFA in dichloromethane (2 × 2 h). HPLC purification afforded 8.9 mg **107** (16.3 µmol, 27% yield over 14 steps).

MALDI-TOF (matrix: DHB/CHCA 1:1, 0.1% TFA) m/z: 547 $[M+H]^+$, 569 $[M+Na]^+$.– Analytical HPLC (5–95% acetonitrile in water + 0.1% TFA over 20 min): t_R = 10.8 min, 96% purity.

H-(Sar-N2mo)$_3$-OH (109)

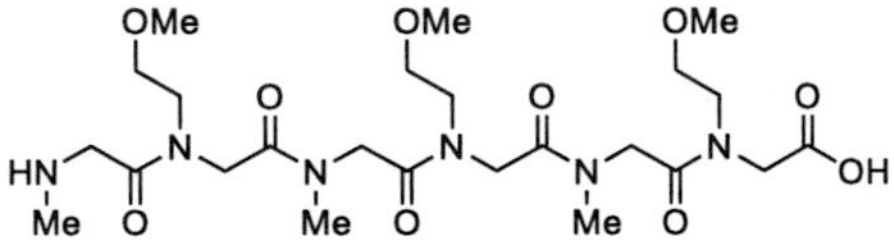

OMe OMe OMe
O O O
HN N N N N OH
Me O Me O Me O

105 mg 2-chlorotrityl chloride resin (0.168 mmol, 1.00 equiv.) were treated as described in **GP8**. For the first reaction 30 mg bromoacetic acid (0.218 mmol, 1.30 equiv.) and 0.15 mL DIPEA (113 mg, 0.874 mmol, 5.20 equiv.) dissolved in 0.45 mL dichloromethane were used. The mixture was shaken for 45 min. N2mo side-chains were introduced according to **GP6**. Acylations were performed with 184 mg bromoacetic acid (1.33 mmol, 7.90 equiv.) 1 M in DMF and 0.21 mL DIC (167 mg, 1.33 mmol, 7.90 equiv.) and shaken for 45 min. For the substitution reactions 0.15 mL methoxyethylamine (126 mg, 1.68 mmol, 10.0 equiv.) 1 M in DMF were employed and shaken for 1 h. Fmoc-sarcosine (157 mg, 0.504 mmol, 3.00 equiv.), was coupled to the resin as described in **GP5** with 68 mg HOBt·(0.504 mmol, 3.00 equiv.) and 78 µL DIC (64 mg, 0.504 mmol, 3.00 equiv.) in 1 mL DMF. The reactions were shaken for 3 h and overnight (double coupling). Methoxyethylamine and sarcosine couplings were alternated until the desired peptoid length was reached. Finally the peptoid was cleaved from the resin according to **GP13** with 1 mL 20% HFIP in dichloromethane (2 × 2 h). The obtained peptoid **109** was cyclized without further purification.

Peptoid **Fmoc-109**: MALDI-TOF (matrix: DHB/CHCA 1:1, 0.1% TFA) m/z: 821 $[M+Na]^+$, 837 $[M+K]^+$.

c-(Sar-N2mo)$_3$ (110)

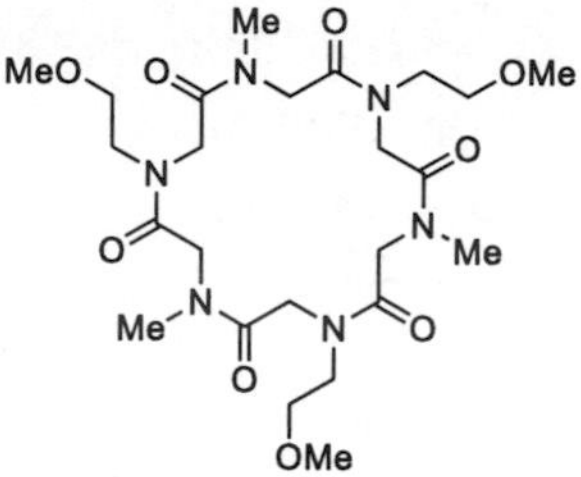

21 mg of crude linear peptoid **109** (36 µmol, 1.00 equiv.) were cyclized following **GP14** with 57 mg PyBOP (0.109 mmol, 3.00 equiv.) and 37 µL DIPEA (28 mg, 0.219 mmol, 6.00 equiv.) in 20 mL dry DMF for 1 h. Finally the peptoid was purified *via* HPLC affording 0.8 mg **110** (1.4 µmol, 0.8% yield over 14 steps).

MALDI-TOF (matrix: DHB/CHCA 1:1, 0.1% TFA) m/z: 581 $[M+Na]^+$.– Analytical HPLC (5–95% acetonitrile in water + 0.1% TFA over 20 min): t_R = 11.8 min, 96% purity.

Ac-N2ad-N4am-N2in-N2im-Sar-NH$_2$ (113)

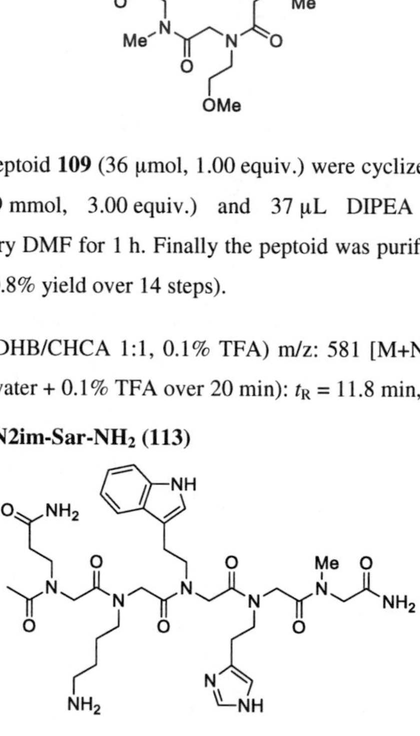

200 mg low-loading Rink-amide resin (0.058 mmol, 1.00 equiv.) were swollen and deprotected following **GP1**. Fmoc-sarcosine (54 mg, 0.174 mmol, 3.00 equiv.) was coupled as described in **GP2** with 27 mg HOBt·H_2O (0.174 mmol, 3.00 equiv.) and 27 µL DIC (22 mg, 0.174 mmol, 3.00 equiv.) 0.2 M in DMF. After Fmoc-deprotection, residue N2im was introduced according to **GP6**. For the acylation step 64 mg bromoacetic acid (0.458 mmol, 7.90 equiv.) 1 M in DMF and 71 µL DIC (58 mg, 0.458 mmol, 7.90 equiv.) were used. The reaction was shaken for 1 h. Substitution was performed with histamine (**41**, 51 mg, 0.458 mmol, 7.90 equiv.) 1 M in DMF and shaken for 1 h 15 min. The following side-chains (N2in, N4am^{Boc} and N2ad) were coupled to the resin according to **GP7**. For the acylation steps 49 mg chloracetic acid (0.522 mmol, 9.00 equiv.) 1 M in DMF and 81 µL DIC (66 mg,

0.522 mmol, 9.00 equiv.) were used. The reactions were shaken for 2 h and 4.5 h (after histamine). Substitutions were performed with tryptamine (**39**, 84 mg, 0.522 mmol, 9.00 equiv.) 1 M in DMF overnight, Boc-protected diaminobutane **81** (98 mg, 0.522 mmol, 9.00 equiv.) 1 M in DMF for 3.5 h, and amine **44** (46 mg, 0.522 mmol, 9.00 equiv.) 1 M in NMP overnight. Finally the peptoid was capped with 55 µL acetic anhydride (59 mg, 0.580 mmol, 10.0 equiv.) and 148 µL DIPEA (112 mg, 0.870 mmol, 15.0 equiv.) in 1 mL DMF for 3.5 h as described in **GP11**. Cleavage from the resin was achieved with 2 mL 95% TFA solution in dichloromethane, as described in **GP12** during 3 and 1.5 h. After HPLC purification 15 mg peptoid **113** (20 µmol, 35% yield over 13 steps) were isolated. The cleanest fraction (2 mg, 2.7 µmol, 98% HPLC purity) was directly used in biological assays. Fractions with slightly lower purity (13 mg, 75–90% HPLC purity) were guanidinated without further purification to yield **Sar-1**.

MALDI-TOF (matrix: DHB/CHCA 1:1, 0.1% TFA) m/z: 738 $[M+H]^+$, 760 $[M+Na]^+$, 776 $[M+K]^+$.– Analytical HPLC (5–95% acetonitrile in water + 0.1% TFA over 20 min): t_R = 8.5 min, 98% purity.

Ac-N2ad-N4gn-N2in-N2im-Sar-NH_2 (Sar-1)

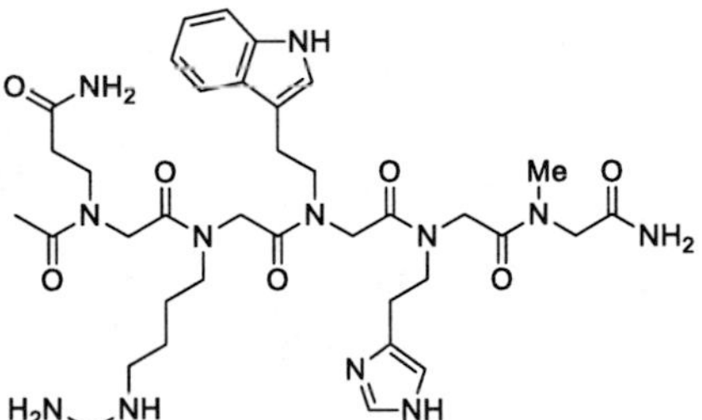

13 mg peptoid **113** (0.018 mmol, 1.00 equiv.) were guanidinated according to **GP10** with 8 mg 1*H*-pyrazole-1-carboxamidine hydrochloride (**84**, 53 µmol, 3.00 equiv.) and 27 µL DIPEA (21 mg, 0.159 mmol, 9.00 equiv.) in 1.8 mL DMA. Finally the peptoid was purified *via* HPLC affording 1.9 mg of **Sar-1** (2.4 µmol, 13% yield).

MALDI-TOF (matrix: DHB/CHCA 1:1, 0.1% TFA) m/z: 780 $[M+H]^+$, 802 $[M+Na]^+$.– Analytical HPLC (5–95% acetonitrile in water + 0.1% TFA over 20 min): t_R = 8.7 min, 99% purity.

Ac-N2ad-N4am-N2in-Sar-N3cx-NH_2

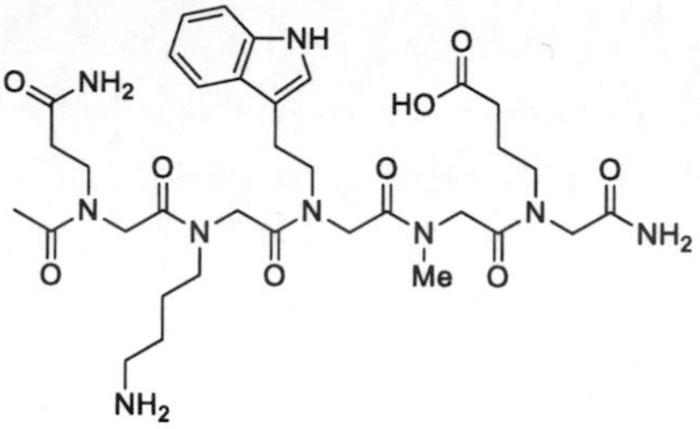

200 mg low-loading Rink-amide resin (0.058 mmol, 1.00 equiv.) were swollen and deprotected following **GP1**. Submonomer N3cx^{tBu} was coupled according to **GP6**. For the acylation step 64 mg bromoacetic acid (0.458 mmol, 7.90 equiv.) 1 M in DMF and 0.1 mL DIC (81 mg, 0.642 mmol, 11.1 equiv.) were used. The reaction was shaken for 1 h. Substitution was performed with 65 mg amine **43** (0.406 mmol, 7.00 equiv.) 1 M in DMF overnight. Fmoc-sarcosine (54 mg, 0.174 mmol, 3.00 equiv.) was coupled as described in **GP2** with 27 mg HOBt·H_2O (0.174 mmol, 3.00 equiv.) and 27 µL DIC (22 mg, 0.174 mmol, 3.00 equiv.) 0.2 M in DMF. After Fmoc-deprotection, submonomers N2in, N4am^{Boc} and N2ad were built in parallel to peptoid **113** using the same procedures, reaction times and amount of reagents. Capping and cleavage from the resin were also performed as described for **113**. HPLC purification afforded 6 mg peptoid Ac-N2ad-N4am-N2in-Sar-N3cx-NH_2 (8.2 µmol, 14% yield over 13 steps) that were guanidinated without further purification to yield **Sar-2**.

MALDI-TOF (matrix: DHB/CHCA 1:1, 0.1% TFA) m/z: 730 $[M+H]^+$, 752 $[M+Na]^+$, 768 $[M+K]^+$.– Analytical HPLC (5–95% acetonitrile in water + 0.1% TFA over 20 min): t_R = 8.7 min, 78–88% purity.

Ac-N2ad-N4gn-N2in-Sar-N3cx-NH$_2$ (Sar-2)

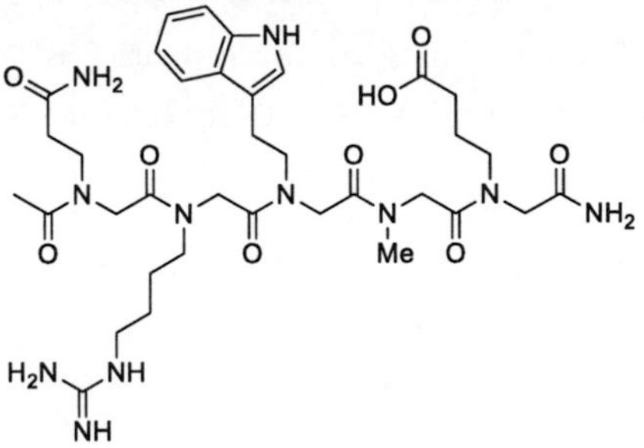

6 mg peptoid Ac-N2ad-N4am-N2in-Sar-N3cx-NH$_2$ (8.2 µmol, 1.00 equiv.) were guanidinated according to **GP10** with 4 mg 1*H*-pyrazole-1-carboxamidine hydrochloride (**84**, 25 µmol, 3.00 equiv.) and 13 µL DIPEA (10 mg, 74 µmol, 9.00 equiv.) in 1.4 mL DMA. Finally the peptoid was purified *via* HPLC obtaining 1.9 mg of **Sar-2** (2.5 µmol, 30% guanidination yield, 4.3% yield over 14 steps).

MALDI-TOF (matrix: DHB/CHCA, 0.1% TFA) m/z: 772 $[M+H]^+$.– Analytical HPLC (5–95% acetonitrile in water + 0.1% TFA over 20 min): $t_R = 9.1$ min, 90% purity.

Ac-N2ad-N4am-Sar-N2im-N3cx-NH$_2$

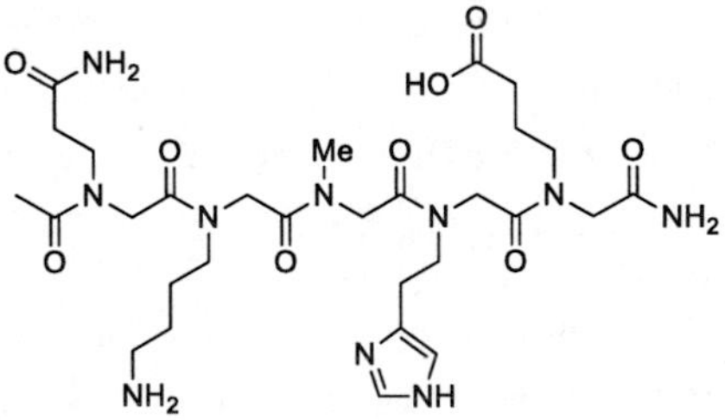

200 mg low-loading Rink-amide resin (0.058 mmol, 1.00 equiv.) were swollen and deprotected following **GP1**. Submonomers N3cx^{tBu} and N2in were coupled according to **GP6**. For the acylation step 64 mg bromoacetic acid (0.458 mmol, 7.90 equiv.) 1 M in DMF and 0.1 mL DIC (81 mg, 0.642 mmol, 11.1 equiv.) were used. Acylations were performed with amine **43** (65 mg, 0.406 mmol, 7.00 equiv.) 1 M in DMF shaken overnight and histamine (**41**, 51 mg, 0.458 mmol, 7.90 equiv.) 1 M in DMF shaken for 1 h 15 min. Then Fmoc-sarcosine (54 mg, 0.174 mmol, 3.00 equiv.) was coupled as described in **GP2** with 27 mg HOBt·H_2O (0.174 mmol, 3.00 equiv.) and 27 µL DIC (22 mg, 0.174 mmol, 3.00 equiv.) 0.2 M in DMF. After Fmoc-deprotection, submonomers N4am^{Boc} and N2ad were built in parallel to the

synthesis of peptoid **113** using the same procedures, reaction times and amount of reagents. Capping and cleavage from the resin were also performed as described for **113**. HPLC purification afforded 1.6 mg peptoid Ac-N2ad-N4am-Sar-N2im-N3cx-NH_2 (2.4 μmol, 4% yield over 13 steps).

MALDI-TOF (matrix: DHB/CHCA 1:1, 0.1% TFA) m/z: 681 $[M+H]^+$, 703 $[M+Na]^+$, 719 $[M+K]^+$. – Analytical HPLC (5–95% acetonitrile in water + 0.1% TFA over 20 min): t_R = 3.8 min, >99% purity.

Ac-N2ad-N4gn-Sar-N2im-N3cx-NH_2 (Sar-3)

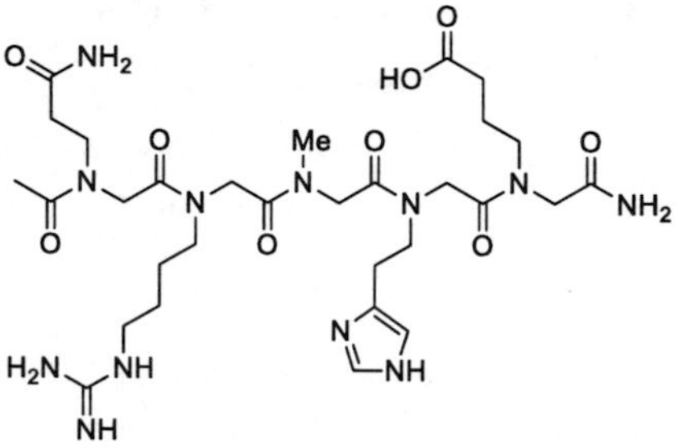

1.6 mg peptoid Ac-N2ad-N4am-Sar-N2im-N3cx-NH_2 (8.2 μmol, 1.00 equiv.) were guanidinated according to **GP10** with 1 mg 1*H*-pyrazole-1-carboxamidine hydrochloride (**84**, 7.1 μmol, 3.00 equiv.) and 4 μL DIPEA (3 mg, 21 μmol, 9.00 equiv.) in 1.8 mL DMA. Finally the peptoid was purified *via* HPLC affording 0.4 mg of **Sar-3** (0.6 μmol, 24% guanidination yield, 1.0% yield over 14 steps).

MALDI-TOF (matrix: DHB/CHCA, 0.1% TFA) m/z: 723 $[M+H]^+$, 745 $[M+Na]^+$, 761 $[M+K]^+$. – Analytical HPLC (5–95% acetonitrile in water + 0.1% TFA over 20 min): t_R = 11.0 min, 95% purity.

Ac-N2ad-Sar-N2in-N2im-N3cx-NH_2 (Sar-4)

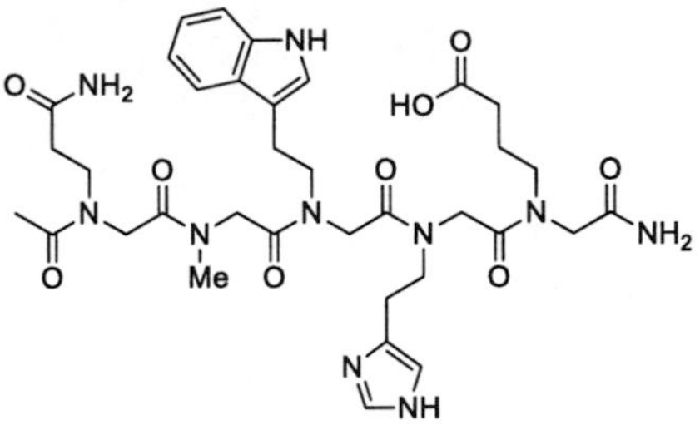

200 mg low-loading Rink-amide resin (0.058 mmol, 1.00 equiv.) were swollen and deprotected following **GP1**. Submonomer N3cx^{tBu} was coupled according to **GP6**. For the acylation step 64 mg bromoacetic acid (0.458 mmol, 7.90 equiv.) 1 M in DMF and 0.1 mL DIC (81 mg, 0.642 mmol, 11.1 equiv.) were used. The reaction was shaken for 1 h. Substitution was performed with 65 mg amine **43** (0.406 mmol, 7.00 equiv.) 1 M in DMF overnight. Then submonomers N2im and N2in were coupled in parallel to the synthesis of peptoid **113** using the same procedures, reaction times and amount of reagents. Fmoc-sarcosine (54 mg, 0.174 mmol, 3.00 equiv.) was coupled as described in **GP2** with 27 mg HOBt·H_2O (0.174 mmol, 3.00 equiv.) and 27 µL DIC (22 mg, 0.174 mmol, 3.00 equiv.) 0.2 M in DMF. After Fmoc-deprotection, incorporation of submonomer N2ad, capping and cleavage from the resin were also performed as described for **113**. HPLC purification afforded 0.5 mg peptoid **Sar-4** (0.66 µmol, 1.1% yield over 13 steps).

MALDI-TOF (matrix: DHB/CHCA 1:1, 0.1% TFA) m/z: 753 $[M+H]^+$, 775 $[M+Na]^+$, 791 $[M+K]^+$.– Analytical HPLC (5–95% acetonitrile in water + 0.1% TFA over 20 min): t_R = 8.9 min, 95% purity.

Ac-Sar-N4am-N2in-N2im-N3cx-NH_2

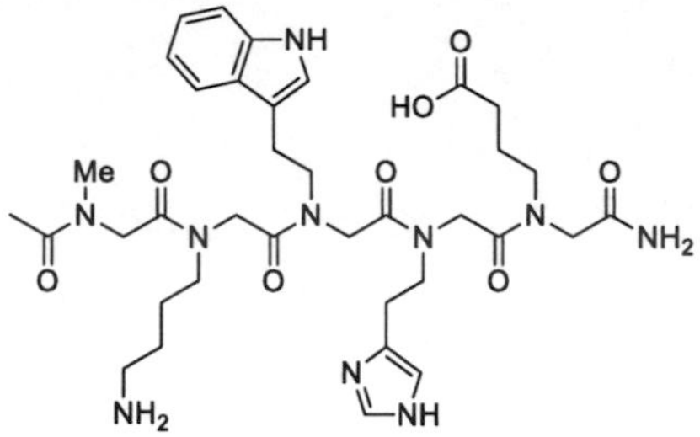

200 mg low-loading Rink-amide resin (0.058 mmol, 1.00 equiv.) were swollen and deprotected following **GP1**. Submonomer $N3cx^{tBu}$ was coupled according to **GP6**. For the acylation step 64 mg bromoacetic acid (0.458 mmol, 7.90 equiv.) 1 M in DMF and 0.1 mL DIC (81 mg, 0.642 mmol, 11.1 equiv.) were used. The reaction was shaken for 1 h. Substitution was performed with 65 mg amine **43** (0.406 mmol, 7.00 equiv.) 1 M in DMF overnight. Then, submonomers N2im, N2in and $N4am^{Boc}$ were coupled in parallel to the synthesis of peptoid **113** using the same procedures, reaction times and amount of reagents. Fmoc-sarcosine (54 mg, 0.174 mmol, 3.00 equiv.) was coupled as described in **GP2** with 27 mg $HOBt \cdot H_2O$ (0.174 mmol, 3.00 equiv.) and 27 µL DIC (22 mg, 0.174 mmol, 3.00 equiv.) 0.2 M in DMF. After Fmoc-deprotection, capping and cleavage from the resin were also performed as described for **113**. HPLC purification afforded 3 mg peptoid Ac-Sar-N4am-N2in-N2im-N3cx-NH_2 (3.9 µmol, 6.7% yield over 13 steps) that were used in biological assays and 49 mg of lower purity fractions (66–85% HPLC purity) that were guanidinated without further purification to yield **Sar-5**.

MALDI-TOF (matrix: DHB/CHCA 1:1, 0.1% TFA) m/z: 753 $[M+H]^+$, 775 $[M+Na]^+$, 791 $[M+K]^+$.– Analytical HPLC (5–95% acetonitrile in water + 0.1% TFA over 20 min): t_R = 8.4 min, >99% purity.

Ac-Sar-N4gn-N2in-N2im-N3cx-NH_2 (Sar-5)

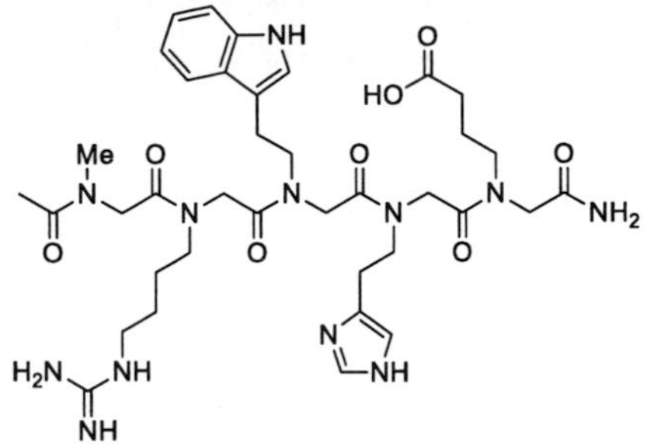

49 mg peptoid Ac-Sar-N4am-N2in-N2im-N3cx-NH_2 (65 μmol, 1.00 equiv.) were guanidinated according to **GP10** with 29 mg 1*H*-pyrazole-1-carboxamidine hydrochloride (0.195 mmol, 3.00 equiv.) and 0.1 mL DIPEA (76 mg, 0.586 mmol, 9.00 equiv.) in 1.8 mL DMA. Finally the peptoid was purified *via* HPLC affording 1.6 mg of **Sar-5** (2.4 μmol, 4.4% yield over 14 steps).

MALDI-TOF (matrix: DHB/CHCA, 0.1% TFA) m/z: 795 $[M+H]^+$, 817 $[M+Na]^+$.– Analytical HPLC (5–95% acetonitrile in water + 0.1% TFA over 20 min): t_R = 8.8 min, 95% purity.

H-N2in-N2im-N3cx^{tBu}-NH-PS (114)

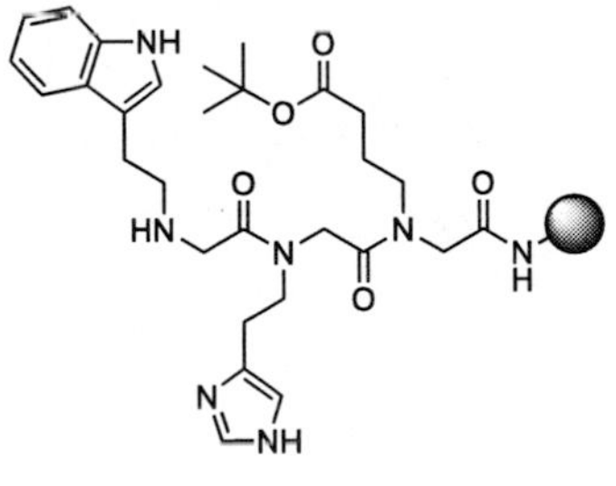

400 mg low-loading Rink-amide resin (0.116 mmol, 1.00 equiv.) were swollen and deprotected following **GP1**. Residues N3cx^{tBu} and N2im were coupled according to **GP3**. The acylations were performed with 127 mg bromoacetic acid (0.916 mmol, 7.90 equiv.) 1 M in DMF and 0.143 mL DIC (115 mg, 0.916 mmol, 7.90 equiv.). For the substitution reactions, tBu-GABA (**43**. 146 mg, 0.92 mmol, 7.90 equiv.) and histamine (**41**, 116 mg, 0.92 mmol, 7.90 equiv.), both as 1 M solution in DMF, were employed. Then submonomer N2in was incorporated as described in **GP4**. The acylation step was performed with 99 mg chloracetic acid (1.04 mmol, 9.00 equiv.) 1 M in DMF and 0.162 mL DIC (132 mg, 1.04 mmol,

9.00 equiv.). For the substitution reaction tryptamine (**39**, 167 mg, 1.04 mmol, 9.00 equiv.) 1 M in DMF was used. Finally, the resin was washed with dichloromethane (3 × 5 mL) and dried overnight.

MALDI-TOF (matrix: DHB/CHCA, 0.1% TFA) m/z: 512.1 $[M+H]^+$.

H-Sar-N4gn-N2in-N2im-N3cx-NH_2 (115)

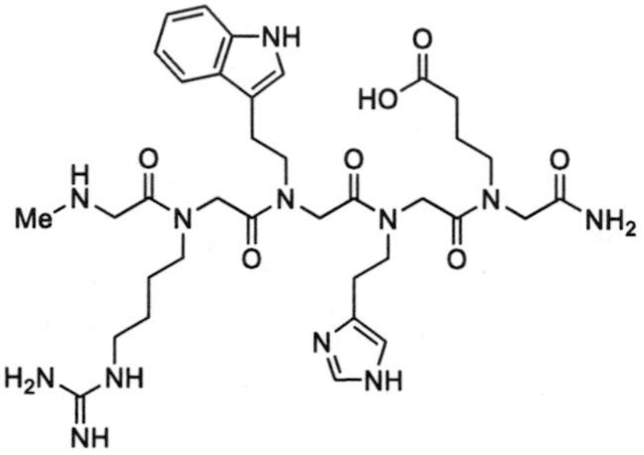

236 mg of the resin-bound peptoid **114** (61.0 µmol, 1.00 equiv.) were swollen in 2 mL DMF for 2 h. Submonomer $N4gn^{2\times Boc}$ was incorporated as described in **GP4**. The acylation step was performed with 52.0 mg chloracetic acid (0.552 mmol, 9.00 equiv.) 1 M in DMF and 0.086 mL DIC (70.0 mg, 0.552 mmol, 9.00 equiv.). For the substitution amine **42** (182 mg, 0.552 mmol, 9.00 equiv.) 1 M in DMF was used. Fmoc-Sarcosine (57.0 mg, 0.184 mmol, 3.00 equiv.) 0.2 M in DMF, was coupled to the resin according to **GP2** with 25 mg HOBt (0.184 mmol, 3.00 equiv.) and 29 µL DIC (23 mg, 0.184 mmol, 3.00 equiv.). Finally the peptoid was cleaved from the resin with 2 mL 95%TFA solution in dichloromethane, as described in **GP12** (2 × 2 h). HPLC purification afforded 1.8 mg peptoid **115** (2.39 µmol, 3.9% yield).

MALDI-TOF (matrix: DHB/CHCA, 0.1 % TFA) m/z: 754 $[M+H]^+$. – Analytical HPLC (5–95% acetonitrile in water + 0.1%TFA over 20 min): t_R = 10.2 min, 93–95% purity.

Ac-Gly-N4gn-N2in-N2im-N3cx-NH_2 (117)

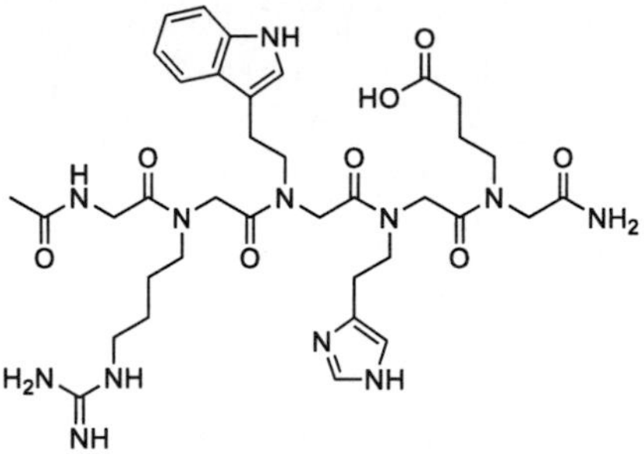

225 mg of the resin bound peptoid **114** (59.0 µmol, 1.00 equiv.) were swollen in 2 mL DMF for 2 h. Submonomer N4am^{Boc} was incorporated as described in **GP4**. The acylation was performed with 50 mg chloracetic acid (0.527 mmol, 9.00 equiv.) 1 M in DMF and 82 µL DIC (66.0 mg, 0.527 mmol, 9.00 equiv.). For the substitution, Boc-protected diaminobutane **81** (99 mg, 0.527 mmol, 9.00 equiv.) 1 M in DMF was used. Fmoc-glycine (52.0 mg, 0.176 mmol, 3.00 equiv.) 0.2 M in DMF was coupled to the resin according to **GP2** with 24 mg HOBt (0.176 mmol, 3.00 equiv.) and 27 µL DIC (22 mg, 0.176 mmol, 3.00 equiv.). After Fmoc-deprotection, the peptoid was capped with 55 µL acetic anhydride (60 mg, 0.585 mmol, 10.0 equiv.) and 149 µL DIPEA (113 mg, 0.878 mmol, 15.0 equiv.) in 0.7 mL DMF overnight as described in **GP11**. Finally the peptoid was cleaved from the resin with 2 mL 95%TFA solution in dichloromethane, as described in **GP12** (2 × 2 h). Guanidination of the amine side-chain was carried out in solution according to **GP10** with 26.0 mg 1*H*-pyrazole-1-carboxamidine monohydrochloride (0.177 mmol, 3.00 equiv.) and 0.091 mL DIPEA (69.0 mg, 0.531 mmol, 9.00 equiv.) in 2.0 mL DMA. HPLC purification afforded 1.2 mg peptoid **117** (1.54 µmol, 2.6% yield).

MALDI-TOF (matrix: DHB/CHCA, 0.1 % TFA) m/z: 782 $[M+H]^+$.– Analytical HPLC (5–95% acetonitrile in water + 0.1%TFA over 20 min): t_R = 10.8 min, 95% purity.

H-N4gn-N32m-N3sm-N43m-N42m-Gly-N5am-NH_2 (120)

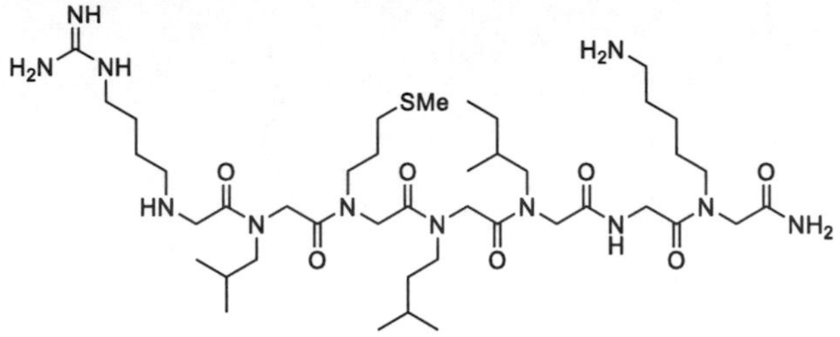

211 mg Rink-amide resin (0.140 µmol, 1.00 equiv.) were swollen and deprotected following **GP1**. Fmoc-N5am^{Boc}-OH (52.0 mg, 0.176 mmol, 3.00 equiv.), available in the group, and Fmoc-Gly-OH (52.0 mg, 0.176 mmol, 3.00 equiv.) were coupled to the resin according to **GP2.** For each coupling 67 mg HOBt (0.440 mmol, 3.00 equiv.) and 0.11 mL DIC (90 mg, 0.71 mmol, 5.00 equiv.) dissolved in 3 mL DMF were used. The following side-chains (N42m, N43m, N3sm, N32m and N4gn$^{2\times Boc}$) were all introduced as submonomers as described in **GP3**. The acylations were performed with 150 mg bromoacetic acid (1.11 mmol, 7.90 equiv.) and 0.28 mL DIC (0.23 g, 1.83 mmol, 13.0 equiv.) dissolved in 1 mL DMF. For the substitution reactions the following amines were employed: 2-methylbutan-1-amine (**121**, 140 µL, 100 mg, 1.15 mmol, 7.90 equiv.), 3-methylbutan-1-amine (**122**, 130 µL, 100 mg, 1.15 mmol, 7.90 equiv.), 3-(methylthio)propane-1-amine (**123**, 130 µL, 120 mg, 1.15 mmol, 7.90 equiv.), isobutylamine (**124**, 110 µL, 84 mg, 1.15 mmol, 7.90 equiv.) and amine **42** (380 mg, 1.15 mmol, 7.90 equiv.), each dissolved in 2 mL DMF. Finally the peptoid was cleaved from the resin with 2 mL 95% TFA solution in dichloromethane overnight, as described in **GP12**. HPLC purification afforded 34 mg peptoid **120** (17.6 µmol, 12% yield over 16 steps).

MALDI-TOF (matrix: DHB, 0.1 % TFA) m/z: 899 $[M+H]^+$.

H-N4am-N32m-N3sm-N43m-N42m-Gly-N4am-NH-SP (136)

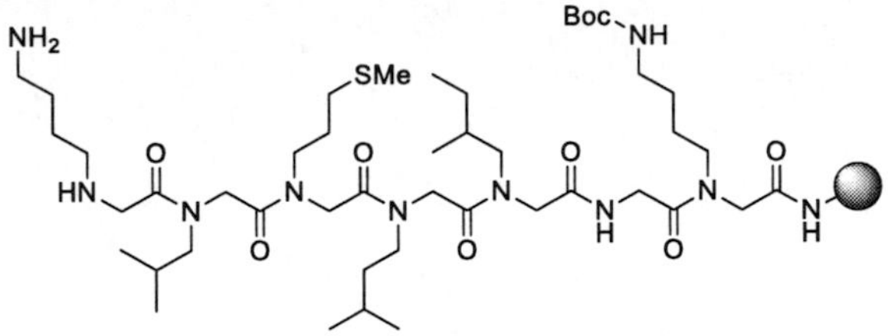

100 mg Rink-amide resin (70.0 μmol, 1.00 equiv.) were swollen and deprotected following **GP1**. Submonomer N4am^{Boc} was introduced according to **GP6**. All side-chains were introduced as submonomers according to **GP6**. The acylation step was performed with 88 mg bromoacetic acid (0.630 mmol, 9.00 equiv.) 1 M in DMF and 76 μL DIC (62 mg, 0.490 mmol, 7.00 equiv.) and was shaken for 20 min. The substitution was performed with Boc-protected diaminobutane **81** (109 mg, 0.581 mmol, 8.30 equiv.) 1 M in DMF and shaken for 2 h. Then, Fmoc-glycine (103 mg, 0.345 mmol, 5.00 equiv.) was coupled to the resin overnight as described in **GP5** with 53 mg HOBt·H_2O (0.345 mmol, 5.00 equiv.) and 0.05 mL DIC (43 mg, 0.345 mmol, 5.00 equiv.) in 3.4 mL DMF. The following submonomers (N42m, N43m, N3sm, N32m and N4am) were also incorporated following **GP6**. The acylation steps were performed as described above. For the substitution reactions the following amines (as 1M solution in DMF) and reaction times were employed: 2-methylbutan-1-amine (**121**, 70 μL, 51 mg, 0.581 mmol, 8.3 equiv., 1 h), 3-methylbutan-1-amine (**122**, 70 μL, 51 mg, 0.581 mmol, 8.3 equiv., 30 min), 3-(methylthio)propane-1-amine (**123**, 60 μL, 61 mg, 0.581 mmol, 8.3 equiv., 30 min) isobutylamine (**124**, 60 μL, 42 mg, 0.581 mmol, 8.3 equiv., 1.5 h) and 1,4-diaminobutane (**51**, 51 mg, 0.581 mmol, 8.30 equiv., 1 h). Then, the resin was dried and divided between peptoid **129** and **130**.

H-N4gn-N32m-N3sm-N43m-N42m-Gly-N4am-NH_2 (129)

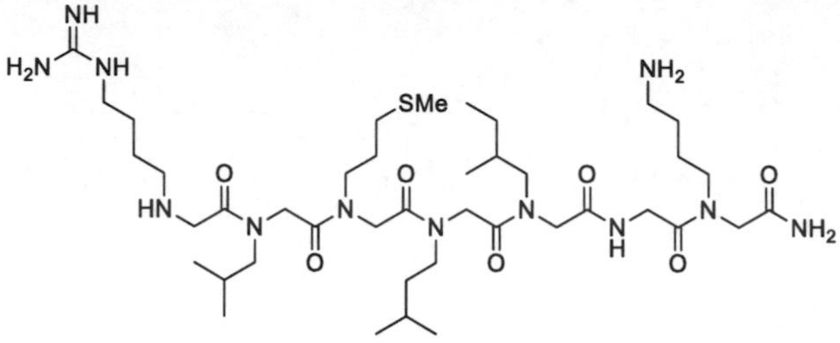

Half of resin bound peptoid **136** (35 µmol, 1.00 equiv.) was guanidinated on solid phase overnight according to **GP9** using 26 mg 1*H*-pyrazole-1-carboxamidine hydrochloride (0.175 mmol, 5.00 equiv.) and 0.09 mL DIPEA (0.525 mmol, 15.00 equiv.) dissolved in 1 mL DMF. Finally the peptoid was cleaved from the resin with 2 mL 95% TFA solution in dichloromethane, as described in **GP12**, for 4 and 2 h. HPLC purification afforded 13 mg peptoid **129** (14.8 µmol, 42% yield over 17 steps) and 2.3 mg dimer **138** (1.4 µmol, 8% yield).

Peptoid **129**: MALDI-TOF (matrix: DHB, 0.1% TFA) m/z: 885 $[M+H]^+$.
Dimer **138**: MALDI-TOF (matrix: DHB, 0.1% TFA) m/z: 1640 ($[M+H]^+$.

Ac-N4gn-N32m-N3sm-N43m-N42m-Gly-N4am-NH_2 (130)

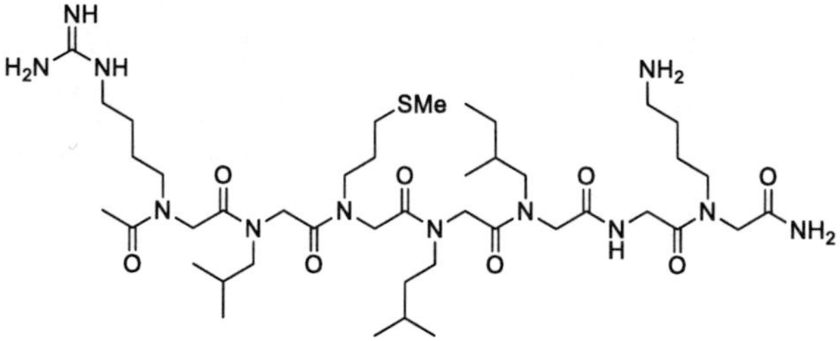

Half of resin-bound peptoid **136** (35 µmol, 1.00 equiv.) was guanidinated on solid phase overnight according to **GP9** using 54 mg *N,N'*-Di-Boc-1*H*-pyrazole-1-carboxamidine (0.175 mmol, 5.00 equiv.) and 0.09 mL DIPEA (0.525 mmol, 15.00 equiv.) dissolved in 1 mL DMF. Then the peptoid was capped overnight following **GP11** with 33 µL acetic anhydride (36 mg, 0.350 mmol, 10.0 equiv.) and 89 µL DIPEA (68 mg, 0.525 mmol, 15.0 equiv.). in 1 mL DMF. Finally the peptoid was cleaved from the resin with 2 mL 95% TFA solution in dichloromethane, as described in **GP12**, for 5 h and 1 h 45 min. HPLC purification afforded

10.8 mg peptoid **130** (11.7 µmol, 33% yield over 18 steps), 2.7 mg dimer **138** (1.7 µmol, 9.7% yield), 2.4 mg dimer **139** (1.4 µmol, 8% yield).

Peptoid **129**: MALDI-TOF (matrix: DHB, 0.1% TFA) m/z: 927 $[M+H]^+$.
Dimer **138**: MALDI-TOF (matrix: DHB, 0.1% TFA) m/z: 1640 ($[M+H]^+$.
Dimer **139**: MALDI-TOF (matrix: DHB, 0.1% TFA) m/z: 1683 ($[M+H]^+$.

H-N4gn-N1ch-N3sm-N43m-N1ch-Gly-N5am-NH₂ (131)

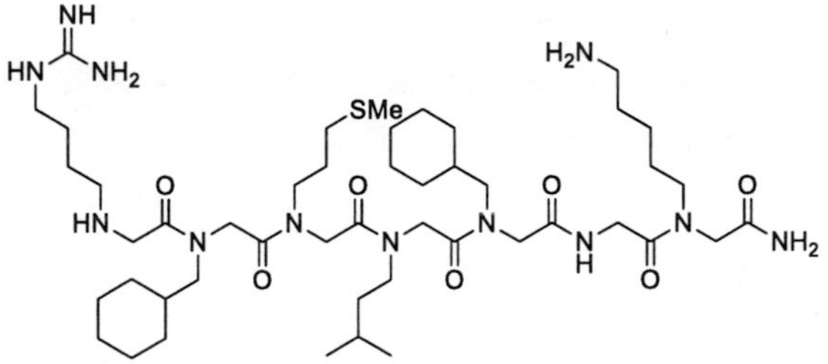

176 mg Rink-amide resin (0.121 mmol, 1.00 equiv.) were swollen and deprotected following **GP1**. Monomer Fmoc-N5am^{Boc}-OH (176 mg, 0.364 mmol, 3.00 equiv.) and Fmoc-glycine (108 mg, 0.364 mmol, 3.00 equiv.) were consecutively coupled to the resin as described in **GP2**. For each coupling 56 mg HOBt·H_2O (0.364 mmol, 3.00 equiv.), 0.09 mL DIC (75 mg, 0.594 mmol, 4.91 equiv.) and 3.6 mL DMF were used. The following side-chains (N1ch, N43m, N3sm, N1ch and N4gn$^{2\times Boc}$) were introduced as submonomers according to **GP3**. The acylations were performed with 133 mg bromoacetic acid (0.959 mmol, 7.90 equiv.) 1 M in DMF and 0.24 mL DIC (198 mg, 1.56 mmol, 13.0 equiv.). For the substitution reactions the following amines were employed: cyclohexylmethanamine (0.12 mL, 109 mg, 0.959 mmol, 7.90 equiv.), 3-methylbutan-1-amine (**122**, 0.11 mL, 84 mg, 0.959 mmol, 7.90 equiv.), 3-(methylthio)propane-1-amine (**123**, 0.11 mL, 101 mg, 0.959 mmol, 7.90 equiv.), again cyclohexylmethanamine (0.12 mL, 109 mg, 0.959 mmol, 7.90 equiv.) and amine **42** (317mg, 0.959 mmol, 7.90 equiv.), each as 1M solution in DMF. Finally the peptoid was cleaved from the resin with 2 mL 95% TFA solution in dichloromethane overnight, as described in **GP12**. HPLC purification afforded 34 mg peptoid **131** (35.4 µmol, 29% yield over 16 steps).

MALDI-TOF (matrix: DHB, 0.1 % TFA) m/z: 965 $[M+H]^+$.

H-N2th-N4gn-N32m-N3sm-N43m-N42m-Gly-NH_2 (132)

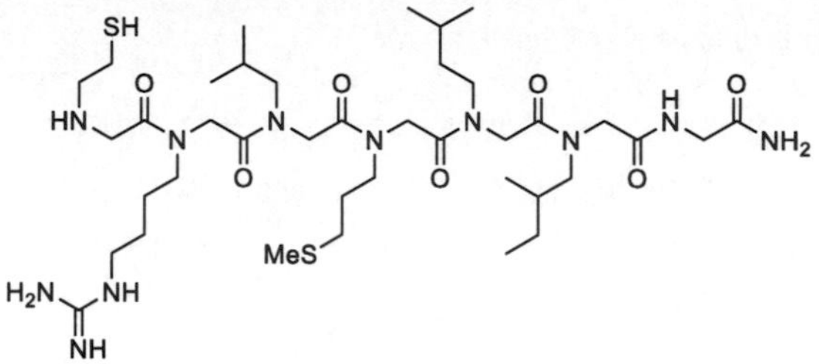

212 mg Rink-amide resin (0.146 μmol, 1.00 equiv.) were swollen and deprotected following **GP1**. Fmoc-Gly-OH (52.0 mg, 0.176 mmol, 3.00 equiv.) was coupled to the resin according to **GP2** with 67 mg HOBt (0.439 mmol, 3.00 equiv.) and 0.11 mL DIC (90 mg, 0.71 mmol, 5.00 equiv.) dissolved in 3 mL DMF. The following side-chains (N42m, N43m, N3sm, N32m and N4gn$^{2\times Boc}$) were all introduced as submonomers as described in **GP3**. The acylations were performed with 161 mg bromoacetic acid (1.16 mmol, 7.90 equiv.) and 0.29 mL DIC (0.24 g, 1.90 mmol, 13.0 equiv.) dissolved in 1 mL DMF. For the substitution reactions the following amines were employed: 2-methylbutan-1-amine (**121**, 140 μL, 101 mg, 1.16 mmol, 7.90 equiv.), 3-methylbutan-1-amine (**122**, 130 μL, 101 mg, 1.16 mmol, 7.90 equiv.), 3-(methylthio)propane-1-amine (**123**, 130 μL, 121 mg, 1.16 mmol, 7.90 equiv.), isobutylamine (**124**, 110 μL, 84 mg, 1.16 mmol, 7.90 equiv.) and amine **42** (382 mg, 1.16 mmol, 7.90 equiv.), each dissolved in 2 mL DMF. Then the resin was dried. ¼ of the resin was separated and used for test sequences. The rest was elongated with submonomer N2th. The acylation was performed with 110 mg bromoacetic acid (0.82 mmol, 7.90 equiv.) and 0.21 mL DIC (0.17 g, 1.34 mmol, 12.9 equiv.) dissolved in 1 mL DMF. For the substitution amine **128** (285 mg, 0.82 mmol, 7.90 equiv.) was employed. Finally the peptoid was cleaved from the resin with 2 mL 95% TFA solution in dichloromethane overnight, as described in **GP12**. HPLC purification afforded 1.4 mg peptoid **132** (1.6 μmol, 1.5% yield over 16 steps).

MALDI-TOF (matrix: DHB, 0.1 % TFA) m/z: 875 $[M+H]^+$.

Rho-N4gn-N32m-N3sm-N43m-N42m-Gly-N5am-NH$_2$ (133)

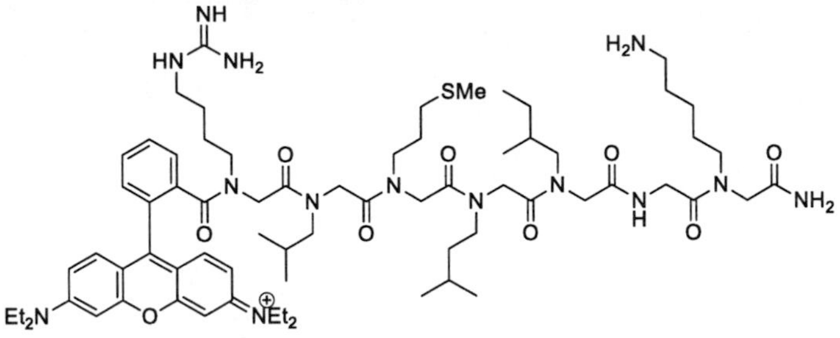

200 mg low-loading Rink-amide resin (68 μmol, 1.00 equiv.) were swollen and deprotected following **GP1**. Residue N5am^{Boc} was coupled to the resin according to **GP6**. The acylation step was performed with 85 mg bromoacetic acid (0.612 mmol, 9.00 equiv.) 1.2 M in DMF and 74 μL DIC (60 mg, 0.476 mmol, 7.00 equiv.) and was shaken for 1 h. For the substitution Boc-protected pentylamine **125** (114 mg, 0.564 mmol, 8.30 equiv.) 1M in DMF was added to the resin and shaken for 1.5 h. Then, Fmoc-glycine (61 mg, 0.204 mmol, 3.00 equiv.) was coupled to the resin as described in **GP5** with 31 mg HOBt·H_2O (0.204 mmol, 3.00 equiv.) and 32 μL DIC (26 mg, 0.204 mmol, 3.00 equiv.) in 1 mL DMF for 1.5 h and overnight (double coupling). After Fmoc-deprotection side-chains N42m, N43m, and N3sm were introduced according to **GP6**. The acylation steps were performed as described above. For the substitution reactions the following amines (as 1 M solution in DMF) and reaction times were employed: 2-methylbutan-1-amine (**121**, 67 μL, 49 mg, 0.564 mmol, 8.3 equiv., 1 h), 3-methylbutan-1-amine (**122**, 66 μL, 49 mg, 0.564 mmol, 8.3 equiv., 1.5 h), and 3-(methylthio)propane-1-amine (**123**, 62 μL, 59 mg, 0.564 mmol, 8.3 equiv., 1.5 h). After submonomer N3sm, residue N32m was incorporated as described in **GP7**. The acylation step was performed with 58 mg chloracetic acid (0.612 mmol, 9.00 equiv.) 1.2 M in DMF and 74 μL DIC (60 mg, 0.476 mmol, 7.00 equiv.) and shaken for 45 min. For the substitution reaction isobutylamine (**124**, 56 μL, 41 mg, 0.564 mmol, 8.3 equiv.,) 1 M in DMF was added to the resin and shaken for 2 h. Then, submonomer N4gn$^{2\times Boc}$ was coupled following **GP6**. The acylation step was performed as described above. For the substitution reaction amine **42** (186 mg, 0.564 mmol, 8.30 equiv.) was added to the resin and shaken overnight. Finally 98 mg Rhodamine B (0.204 mmol, 3.00 equiv.), 31 mg HOBt·H_2O (0.204 mmol, 3.00 equiv.) and 32 μL DIC (26 mg, 0.204 mmol, 3.00 equiv.) dissolved in 1 mL DMF were added to the resin and shaken overnight. Then the resin was washed extensively with DMF until the

washes were colorless and the peptoid was cleaved from the resin with 2 mL 95% TFA solution in dichloromethane, as described in **GP12** (2 × 2 h). HPLC purification afforded 8.9 mg peptoid **133** (6.7 μmol, 10% yield over 16 steps) as a pink powder.

MALDI-TOF (matrix: DHB/CHCA 1:1, 0.1% TFA) m/z: 1324 $[M+H]^+$.– Analytical HPLC (5–95% acetonitrile in water + 0.1% TFA over 20 min): t_R = 14.5 min, >99% purity.

FITC-Ahx-N4gn-N32m-N3sm-N43m-N42m-Gly-N5am-NH₂ (146)

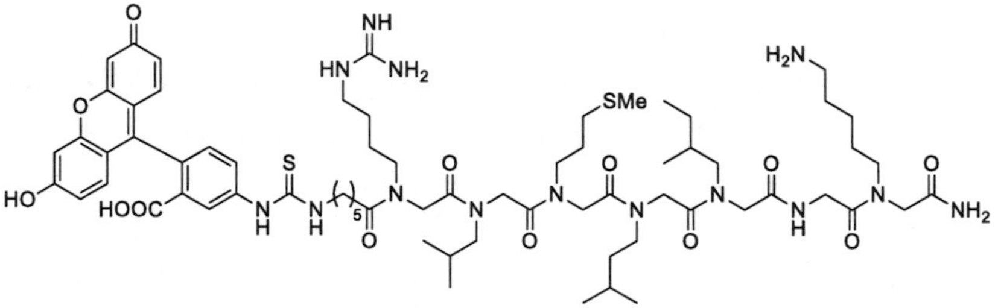

203 mg low-loading Rink-amide resin (59 μmol, 1.00 equiv.) were swollen and deprotected following **GP1**. Residue N5am^{Boc} was coupled to the resin according to **GP6**. The acylation step was performed with 74 mg bromoacetic acid (0.530 mmol, 9.00 equiv.) 1 M in DMF and 64 μL DIC (52 mg, 0.412 mmol, 7.00 equiv.) and shaken for 30 min. For the substitution reaction Boc-protected pentylamine **125** (99 mg, 0.489 mmol, 8.30 equiv.) 1M in DMF was added to the resin and shaken for 2 h. Then, Fmoc-glycine (53 mg, 0.177 mmol, 3.00 equiv.) was coupled to the resin as described in **GP5** with 27 mg HOBt·(0.200 mmol, 3.30 equiv.) and 27 μL DIC (22 mg, 0.177 mmol, 3.00 equiv.) in 0.9 mL DMF for 2 h and overnight (double coupling). After Fmoc-deprotection side-chains N42m, N43m, and N3sm were introduced as submonomers according to **GP6**. The acylation steps were performed as described above. For the substitution reactions the following amines (as 1 M solution in DMF) and reaction times were employed: 2-methylbutan-1-amine (**121**, 58 μL, 43 mg, 0.489 mmol, 8.3 equiv., 45 min), 3-methylbutan-1-amine (**122**, 57 μL, 43 mg, 0.489 mmol, 8.3 equiv., overnight), and 3-(methylthio)propane-1-amine (**123**, 53 μL, 51 mg, 0.489 mmol, 8.3 equiv., 40 min). After submonomer N3sm, residue N32m was incorporated as described in **GP7**. The acylation step was performed with 50 mg chloracetic acid (0.530 mmol, 9.00 equiv.) 1 M in DMF and 64 μL DIC (52 mg, 0.412 mmol, 7.00 equiv.) and shaken for 1 h. For the substitution reaction isobutylamine (**124**, 49 μL, 36 mg, 0.489 mmol, 8.3 equiv.) was added to the resin and shaken overnight. Then submonomer N4gn$^{2×Boc}$ was introduced following **GP6**.

The acylation step was performed as described above. For the substitution reaction amine **42** (161 mg, 0.489 mmol, 8.30 equiv.) was added to the resin and shaken overnight. After that, an Ahx spacer was coupled to the resin following **GP2**. For the coupling 62 mg 6-(Fmoc)aminohexanoic acid (0.177 mmol, 3.00 equiv.) 0.2 M in DMF, 24 mg HOBt·(0.177 mmol, 3.00 equiv.) and 27 μL DIC (22 mg, 0.177 mmol, 3.00 equiv.) were employed. Finally, 76 mg fluorescein isothiocyanate (90% purity, 0.177 mmol, 3.00 equiv.) 1 M in DMF/dichloromehane (1:1) were added to the resin and shaken overnight at room temperature. The procedure was repeated twice. Then the resin was washed extensively with DMF and dichloromethane until the washes were colorless. The peptoid was cleaved from the resin with 2 mL 95% TFA solution in dichloromethane, as described in **GP12** (2 × 2 h). HPLC purification afforded 12.3 mg peptoid **146** (8.8 μmol, 15% yield over 19 steps) as an orange powder.

MALDI-TOF (matrix: DHB/CHCA 1:1, 0.1% TFA) m/z: 1403 $[M+H]^+$, 1425 $[M+Na]^+$. – Analytical HPLC (5–95% acetonitrile in water + 0.1% TFA over 20 min): t_R = 16.7 min, 98% purity.

FITC-Ahx-N4gn-(Sar)$_3$-N42m-Gly-N5am-NH$_2$ (154)

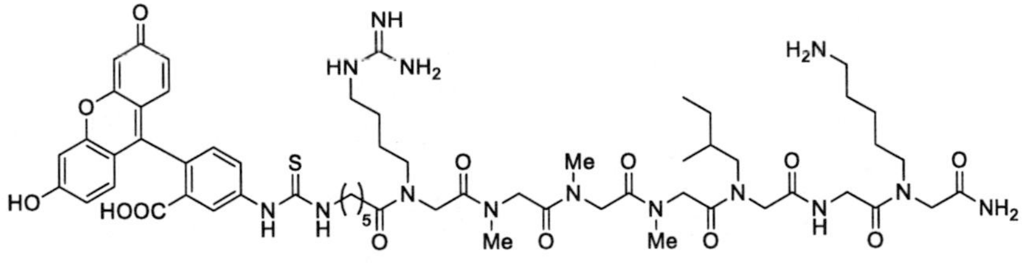

193 mg low-loading Rink-amide resin (56 μmol, 1.00 equiv.) were swollen and deprotected following **GP1**. Residue N5am^{Boc} was coupled to the resin according to **GP3**. The acylation step was performed with 61 mg bromoacetic acid (0.442 mmol, 7.90 equiv.) 0.9 M in DMF and 69 μL DIC (56 mg, 0.442 mmol, 7.90 equiv.). For the substitution reaction Boc-protected pentylamine **125** (89 mg, 0.442 mmol, 7.90 equiv.) 1 M in DMF was used. Then, Fmoc-glycine (50 mg, 0.168 mmol, 3.00 equiv.) was coupled to the resin as described in **GP2** with 23 mg HOBt·(0.168 mmol, 3.00 equiv.) and 26 μL DIC (21 mg, 0.168 mmol, 3.00 equiv.) in 0.8 mL DMF. After Fmoc-deprotection, side-chain N42m was introduced according to **GP3**. The acylation step was performed as described above. For the substitution reaction 2-methylbutan-1-amine (**121**, 53 μL, 39 mg, 0.442 mmol, 7.90 equiv.) 0.9 M in DMF.was

employed. Then, Fmoc-sarcosine (52 mg, 0.168 mmol, 3.00 equiv.) was coupled to the resin three consecutive times, alternating with deprotection steps, according to **GP2** with the same reagents as described above. Side-chain N4gn$^{2\times Boc}$ was also coupled according to **GP3.** The acylation step was performed as described above. For the substitution reaction amine **42** (129 mg, 0.392 mmol, 7.00 equiv.) 0.9 M in DMF was employed. Then 6-(Fmoc)aminohexanoic acid (59 mg, 0.168 mmol, 3.00 equiv.) was coupled to the resin according to **GP2** with the same reagents as described above. Finally, 73 mg fluorescein isothiocyanate (90% purity, 0.168 mmol, 3.00 equiv.) 1 M in DMF/dichloromethane (1:1) were added to the resin and stirred overnight at room temperature. The procedure was repeated twice. Then the resin was washed extensively with DMF and dichloromethane until the washes were colorless. The peptoid was cleaved from the resin with 2 mL 95% TFA solution in dichloromethane, as described in **GP12** for 2 and 3 h. HPLC purification afforded 7.4 mg peptoid **154** (6.0 µmol, 11% yield over 19 steps) as an orange powder.

MALDI-TOF (matrix: DHB/CHCA 1:1, 0.1% TFA) m/z: 1230 $[M+H]^+$, 1463 $[M+Na]^+$. – Analytical HPLC (5–95% acetonitrile in water + 0.1% TFA over 20 min): t_R = 13.3 min, 98% purity.

H-N2mo-Nsmd-(N1ph)$_2$-NH$_2$ (159)

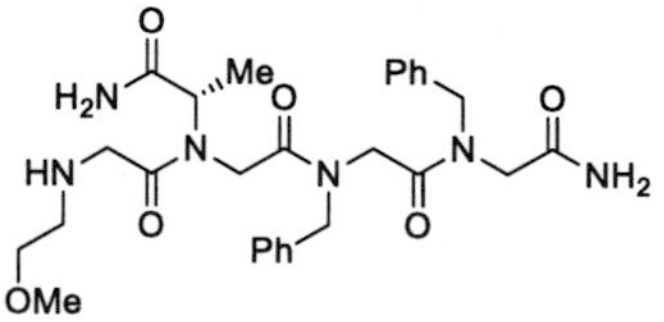

200 mg (66.0 µmol, 1.00 equiv.) of the resin bound peptoid **68** were swollen in 2 mL DMF for 2 h. The peptoid was built as described in **GP3**. For the acylation steps 83 mg bromoacetic acid (0.594 mmol, 9.0 equiv.) 1 M in DMF and 72 µL DIC (58 mg, 0.462 mmol, 7.00 equiv.) were used. Substitutions were performed with 48 mg amine **155** (0.548 mmol, 8.30 equiv.) 1 M in NMP and 48 µL methoxyethylamine (41 mg, 0.548 mmol, 8.30 equiv.) 1 M in DMF. The peptoid was cleaved from the resin with 2 mL 95% TFA in dichloromethane according to **GP12** (2 × 2 h). HPLC purification afforded a fraction of 8.0 mg of an inseparable mixture of **159** and **159-DKP** (14.4 µmol, 22% yield) and another fraction of 9.3 mg **159-DKP** (17.3 µmol, 26% yield). In addition 7.4 mg (16.8 µmol, 26% yield) of the trimer H-Nsmd-(N1ph)$_2$-NH$_2$ could also be isolated.

MALDI-TOF (matrix: DHB, 0.1% TFA) m/z: 555 (**159**, $[M+H]^+$), 538 (**159-DKP**, $[M+H]^+$), 560 (**159-DKP**, $[M+Na]^+$). – Analytical HPLC (5–95% acetonitrile in water + 0.1% TFA over 20 min): t_R = 10.8 min, 69% purity.

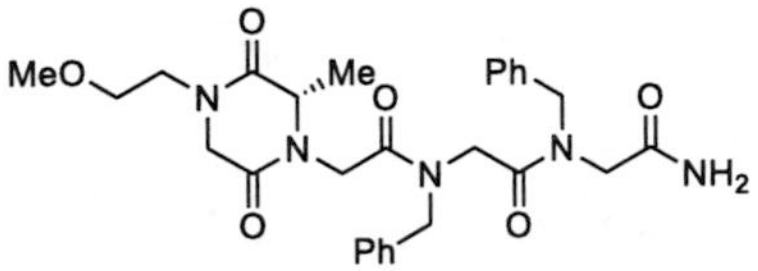

159-DKP

159-DKP: MALDI-TOF (matrix: DHB, 0.1% TFA) m/z: 538 $[M+H]^+$, 560 $[M+Na]^+$. Analytical HPLC (5–95% acetonitrile in water + 0.1% TFA over 20 min): t_R = 11.7 min, 86% purity.

H-Nsmd-(N1ph)₂-NH₂: MS (ESI) m/z: 440 $[M+H]^+$. – Analytical HPLC (5–95% acetonitrile in water + 0.1% TFA over 20 min): t_R = 10.1 min, 95% purity.

H-N2mo-Nrpd-(N1ph)$_2$-NH$_2$ (160)

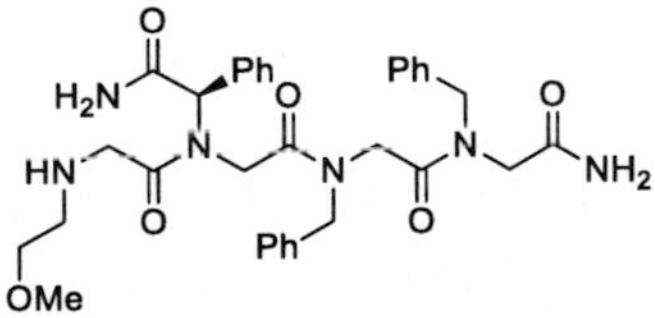

200 mg (66.0 μmol, 1.00 equiv.) of the resin bound peptoid **68** were swollen in 2 mL DMF for 2 h. The peptoid was built as described in **GP3**. For the acylation steps 83 mg bromoacetic acid (0.594 mmol, 9.0 equiv.) 1 M in DMF and 72 μL DIC (58 mg, 0.462 mmol, 7.00 equiv.) were used. Substitutions were performed with 82 mg amine **156** (0.548 mmol, 8.30 equiv.) 1 M in NMP and 48 μL methoxyethylamine (41 mg, 0.548 mmol, 8.30 equiv.) 1 M in DMF. The peptoid was cleaved from the resin with 2 mL 95% TFA in dichloromethane according to **GP12** (2 × 2 h). HPLC purification afforded 16.7 mg of an inseparable mixture of peptoids **160** and **160-DKP** (27.1 μmol, 41% yield).

MALDI-TOF (matrix: DHB, 0.1% TFA) m/z: 617 (**160**, $[M+H]^+$), 600 (**160-DKP**, $[M+H]^+$), 622 (**160-DKP**, $[M+Na]^+$).– Analytical HPLC (5–95% acetonitrile in water + 0.1% TFA over 20 min): t_R = 12.9 min, 94% purity.

H-(Nrpd)$_4$-NH$_2$ (165)

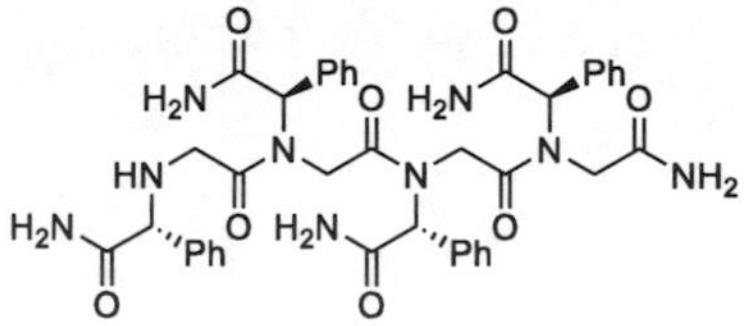

300 mg low-loading Rink-Amide resin (0.102 mmol, 1.00 equiv.) were swollen and deprotected as described in **GP1**. The peptoid was built alternating four acylation and four substitution steps following **GP3**. For the acylation steps 128 mg bromoacetic acid (0.918 mmol, 7.90 equiv.) 1 M in DMF and 0.1 mL DIC (90 mg, 0.714 mmol, 7.00 equiv.) were used. Substitutions were performed with 127 mg amine **156** (0.847 mmol, 8.30 equiv.) 1 M in NMP for 45 min. Acylations after each submonomer were repeated twice (double coupling). After complete assembly of the tetramer, the resin was washed with dichloromethane and dried overnight. Half of the resin (178 mg, 0.051 mmol) was reacted further to form hexamer **167**. The other half was swollen in dichloromethane for 1.5 h and then cleaved from the resin as described in **GP12** with 2 mL 95% TFA solution in dichloromethane (2 × 3 h). HPLC purification afforded 16.1 mg peptoid **165** (20.7 μmol, 41% yield over 10 steps) and 2.1 mg of compound **165-DKP** (2.8 μmol, 5.4% yield).

H-(Nrpd)$_4$-NH$_2$ (165): MALDI-TOF (matrix: DHB/CHCA 1:1, 0.1% TFA) m/z: 799 $[M+Na]^+$, 816 $[M+K]^+$.– Analytical HPLC (5–95% acetonitrile in water + 0.1% TFA over 20 min): t_R = 9.3 min, 97% purity.

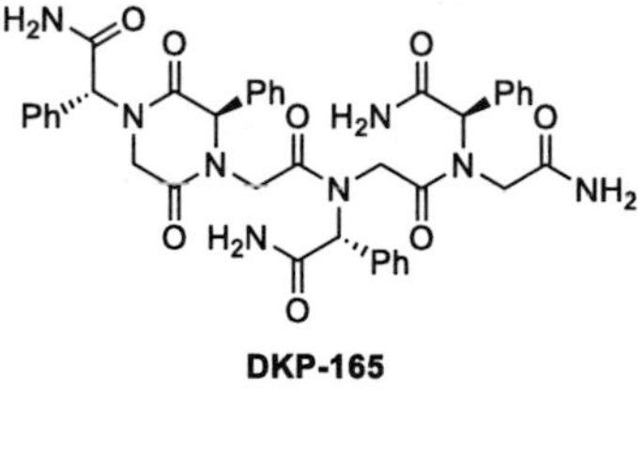

DKP-165

DKP-165: MALDI-TOF (matrix: DHB/CHCA 1:1, 0.1% TFA) m/z: 782 $[M+Na]^+$, 798 $[M+K]^+$.– Analytical HPLC (5–95% acetonitrile in water + 0.1% TFA over 20 min): t_R = 10.3 min, 96% purity.

H-(Nrpd)$_5$-NH$_2$ (166)

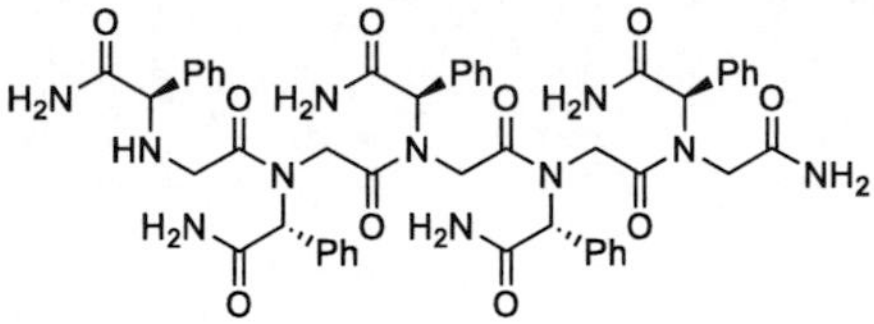

201 mg low-loading Rink-Amide resin (68 µmol, 1.00 equiv.) were swollen and deprotected as described in **GP1**. The peptoid was built alternating five acylation and five substitution steps following **GP3**. For the acylation steps 85 mg bromoacetic acid (0.615 mmol, 9.00 equiv.) 1 M in DMF and 74 µL DIC (60 mg, 0.478 mmol, 7.00 equiv.) were used. Substitutions were performed with 85 mg amine **156** (0.615 mmol, 9.00 equiv.) 1 M in NMP for 45 min. Acylations after each submonomer were repeated twice (double coupling). After complete assembly of the pentamer, the peptoid was cleaved from the resin as described in **GP12** with 2 mL 95% TFA solution in dichloromethane for 3 and 4.5 h. HPLC purification afforded 4.3 mg pentamer **166** (4.4 µmol, 6.5% yield over 12 steps).

MALDI-TOF (matrix: DHB/CHCA 1:1, 0.1% TFA) m/z: 968 $[M+H]^+$, 990 $[M+Na]^+$, 1006 $[M+K]^+$.– Analytical HPLC (5–95% acetonitrile in water + 0.1% TFA over 20 min): t_R = 10.1 min, 94% purity.

H-(Nrpd)$_6$-NH$_2$ (167)

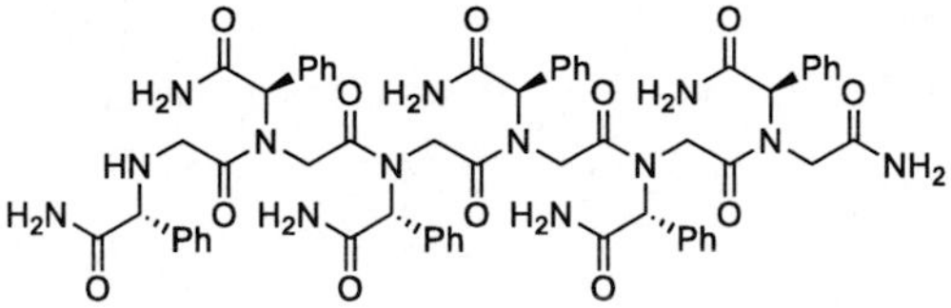

Half of resin-bound tetramer **165** (178 mg, 0.051 mmol, 1.00 equiv.) was reacted twice (double coupling) according to **GP3** with 62 mg bromoacetic acid (0.447 mmol, 9.00 equiv.) 1 M in DMF and 54 µL DIC (44 mg, 0.348 mmol, 7.00 equiv.). The following substitution step was performed with 62 mg amine **156** (0.412 mmol, 8.30 equiv.) 1 M in NMP for 45 min. Repetition of the acylation and substitution steps afforded the hexamer, that was cleaved from the resin as described in **GP12** with 2 mL 95% TFA solution in dichloromethane (2 × 2 h). HPLC purification afforded two fractions 3.7 mg hexamer **167** (3.2 µmol, 6.4% yield over 14 steps). In addition 2.0 mg tetramer **165** (2.6 µmol, 5% yield), 4.5 mg pentamer **166** (4.7 µmol, 9% yield), and 6.4 mg of **166-DKP** byproduct (5.6 µmol, 11% yield) were also isolated.

H-(Nrpd)$_6$-NH$_2$ (167): MALDI-TOF (matrix: DHB/CHCA 1:1, 0.1% TFA) m/z: 1158 $[M+H]^+$, 1180 $[M+Na]^+$, 1196 $[M+K]^+$. – Analytical HPLC (5–95% acetonitrile in water + 0.1% TFA over 20 min): t_R = 10.6 min, 92–97% purity.

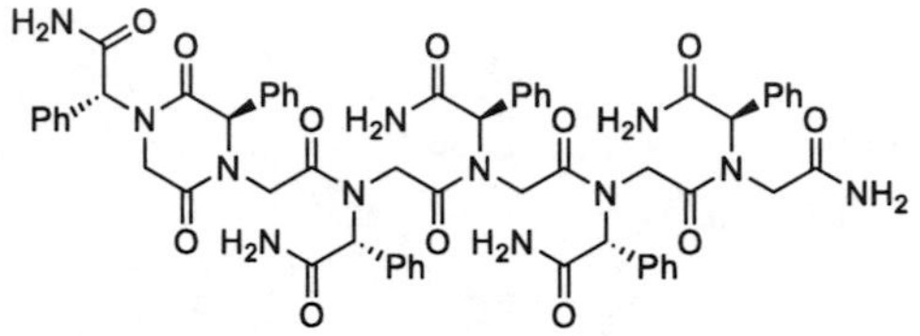

166-DKP

166-DKP: MALDI-TOF (matrix: DHB/CHCA 1:1, 0.1% TFA) m/z: 1162 $[M+Na]^+$, 1178 $[M+K]^+$.– Analytical HPLC (5–95% acetonitrile in water + 0.1% TFA over 20 min): t_R = 11.7 min, 94% purity.

H-(Nrpd)$_5$-NH$_2$ (166): MALDI-TOF (matrix: DHB/CHCA 1:1, 0.1% TFA) m/z: 968 $[M+H]^+$, 990 $[M+Na]^+$, 1006 $[M+K]^+$.– Analytical HPLC (5–95% acetonitrile in water + 0.1% TFA over 20 min): t_R = 10.1 min, 89% purity.

H-(Nrpd)$_4$-NH$_2$ (165): MALDI-TOF (matrix: DHB/CHCA 1:1, 0.1% TFA) m/z: 799 $[M+Na]^+$, 816 $[M+K]^+$.– Analytical HPLC (5–95% acetonitrile in water + 0.1% TFA over 20 min): t_R = 9.3 min, 78% purity.

H-(Nrpd)$_8$-NH$_2$ (168)

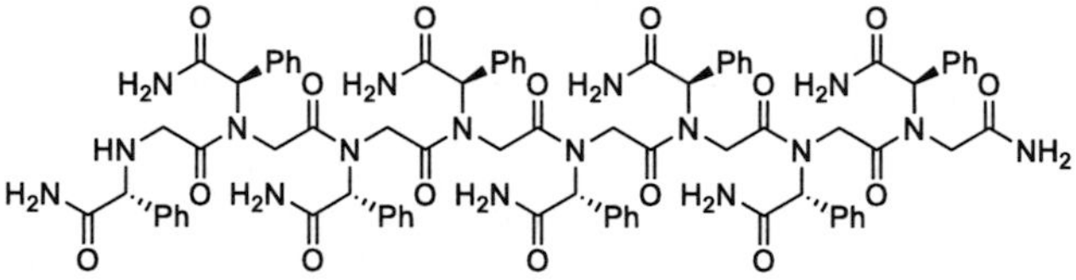

300 mg low-loading Rink-Amide resin (87 µmol, 1.00 equiv.) were swollen and deprotected as described in **GP1**. The peptoid was built alternating acylation and substitution steps (× 8) following **GP3**. For the acylation steps 96 mg bromoacetic acid (0.687 mmol, 7.90 equiv.) 0.9 M in DMF and 0.1 mL DIC (87 mg, 0.687 mmol, 7.90 equiv.) were used. Substitutions were performed with 118 mg amine **156** (0.783 mmol, 9.00 equiv.) 1 M in NMP for 45 min.

Acylations after each submonomer were repeated twice (double coupling). After complete assembly of the octamer, the resin was washed with dichloromethane and dried overnight. Half of the resin (197 mg, 0.041 mmol) was reacted further to form the decamer **169**. The other half was swollen in dichloromethane for 2 h and cleaved as described in **GP12** with 2 mL 95% TFA solution in dichloromethane (2 × 2 h). HPLC purification afforded 1.6 mg octamer **168** (1.0 μmol, 2.4% yield over 18 steps).

MALDI-TOF (matrix: DHB/CHCA 1:1, 0.1% TFA) m/z: 1539 $[M+H]^+$, 1561 $[M+Na]^+$, 1578 $[M+K]^+$.– Analytical HPLC (5–95% acetonitrile in water + 0.1% TFA over 20 min): t_R = 14.8 min, 88% purity.

H-(Nrpd)$_{10}$-NH$_2$ (169)

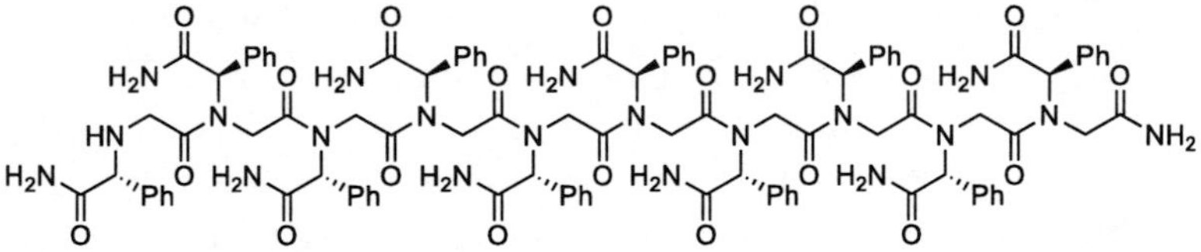

Half of resin-bound octamer **168** (197 mg, 0.041 mmol, 1.00 equiv.) was reacted twice (double coupling) according to **GP3** with 45 mg bromoacetic acid (0.327 mmol, 7.90 equiv.) 1 M in DMF and 51 μL DIC (41 mg, 0.327 mmol, 7.90 equiv.). The following substitution step was performed with 56 mg amine **156** (0.372 mmol, 9.00 equiv.) 1 M in DMF for 45 min. Repetition of the acylation and substitution steps afforded the decamer, that was cleaved from the resin as described in **GP12** with 2 mL 95% TFA solution in dichloromethane (2 × 2 h). HPLC purification afforded 2.3 mg decamer **169** (1.2 μmol, 2.9% yield over 22 steps).

MALDI-TOF (matrix: DHB/CHCA 1:1, 0.1% TFA) m/z: 1920 $[M+H]^+$, 1942 $[M+Na]^+$, 1958 $[M+K]^+$. – Analytical HPLC (5–95% acetonitrile in water + 0.1% TFA over 20 min): t_R = 15.6 min, 81% purity.

H-(Nsmd)$_4$-NH$_2$ (170)

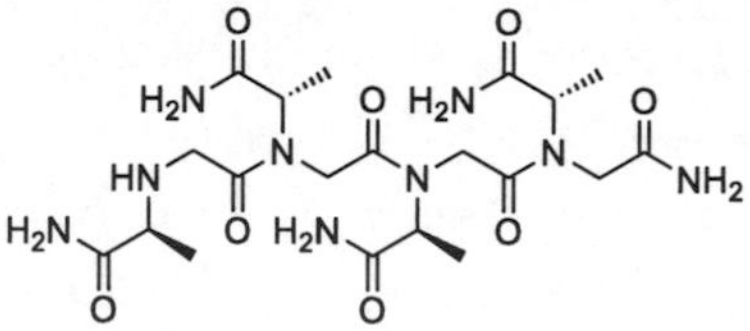

230 mg low-loading Rink-Amide resin (0.078 mmol, 1.00 equiv.) were swollen and deprotected as described in **GP1**. The peptoid was built alternating acylation and substitution steps (× 4) following **GP3**. For each acylation step 27 mg bromoacetic acid (0.196 mmol, 2.50 equiv.) 0.28 M in DMF and 85 μL DIC (69 mg, 0.547 mmol, 7.0 equiv.) were used. Substitutions were performed with 57 mg amine **155** (0.649 mmol, 8.30 equiv.) 1 M in NMP for 45 min. Acylations after each submonomer were repeated twice (double coupling). After assembly of the tetramer, the resin was washed with dichloromethane and dried overnight.

113 mg of dry resin bound tetramer **170** (max. 34.7 μmol, 1.00 equiv.) were swollen in dichloromethane for 1 h and cleaved from the resin as described in **GP12** with 2 mL 95% TFA solution in dichloromethane (2 × 3 h). HPLC purification afforded 1.8 mg peptoid **170** (3.4 μmol, 9.8% yield over 10 steps).

MALDI-TOF (matrix: DHB/CHCA 1:1, 0.1% TFA) m/z: 552 $[M+Na]^+$, 568 $[M+K]^+$.– Analytical HPLC (1–10% acetonitrile in water + 0.1% TFA over 20 min): $t_R = 3.2$ min, 92% purity.

H-(Nsmd)$_5$-NH$_2$ (171)

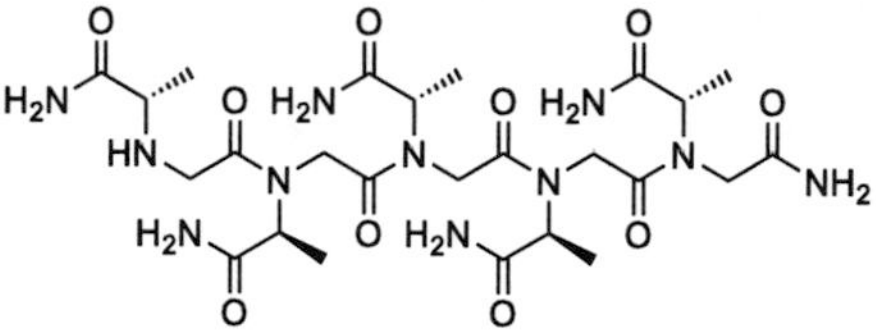

100 mg low-loading Rink-Amide resin (34 μmol, 1.00 equiv.) were swollen and deprotected as described in **GP1**. The peptoid was built following **GP3**. For each acylation step 43 mg bromoacetic acid (0.306 mmol, 9.00 equiv.) 1 M in DMF and 37 μL DIC (30 mg, 0.238 mmol, 7.00 equiv.) were used. Substitutions were performed with 25 mg amine **155** (0.282 mmol, 8.30 equiv.) 1 M in NMP for 45 min. Acylation and substitution reactions were alternated until the pentamer was assembled. Finally the peptoid was cleaved from the resin as described

in **GP12** with 1 mL 95% TFA solution in dichloromethane (2 × 2 h). HPLC purification afforded 0.7 mg **171** (1.1 µmol, 3.1% yield over 12 steps).

MALDI-TOF (matrix: DHB/CHCA 1:1, 0.1% TFA) m/z: 659 $[M+H]^+$, 680 $[M+Na]^+$, 696 $[M+K]^+$. – Analytical HPLC (1–20% acetonitrile in water + 0.1% TFA over 20 min): $t_R = 3.9$ min, 91% purity.

H-(Nsmd)$_6$-NH$_2$ (172)

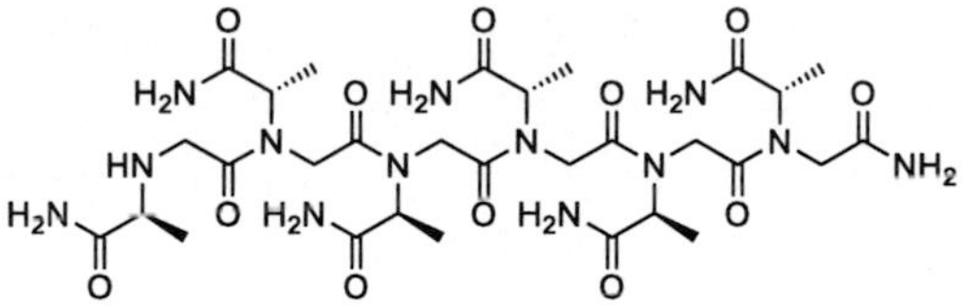

162 mg low-loading Rink-Amide resin (47 µmol, 1.00 equiv.) were swollen and deprotected as described in **GP1**. The peptoid was built alternating acylation and substitution steps following **GP3**. For the acylation steps 52 mg bromoacetic acid (0.371 mmol, 7.90 equiv.) 1 M in DMF and 58 µL DIC (47 mg, 0.371 mmol, 7.90 equiv.) were used. Substitutions were performed with 33 mg amine **155** (0.371 mmol, 7.90 equiv.) 1 M in NMP for 45 min. Acylations after each submonomer were repeated twice (double coupling). After assembly of the hexamer, the resin was divided in two. Half was reacted further to form heptamer **173**. The other half was cleaved as described in **GP12** with 2 mL 95% TFA solution in dichloromethane (2 × 2 h). HPLC purification afforded 3.3 mg **172** (4.2 µmol, 18% yield over 14 steps).

MALDI-TOF (matrix: DHB/CHCA 1:1, 0.1% TFA) m/z: 808 $[M+Na]^+$, 824 $[M+K]^+$. – Semipreparative HPLC (1% acetonitrile in water + 0.1% TFA over 10 min): $t_R = 4.3$ min, 92% purity.

H-(Nsmd)$_n$-NH$_2$ (173)

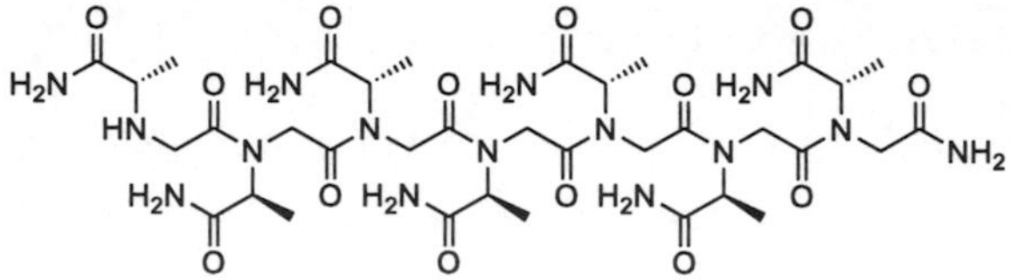

Half of resin-bound hexamer **172** (23 µmol, 1.00 equiv.) was reacted twice (double coupling) with 26 mg bromoacetic acid (0.186 mmol, 7.90 equiv.) 1 M in DMF and 29 µL DIC (23 mg, 0.186 mmol, 7.90 equiv.) following **GP3**. The following substitution step was performed with 16 mg amine **155** (0.186 mmol, 7.90 equiv.) 1 M in NMP for 45 min. Finally the peptoid was cleaved from the resin as described in **GP12** with 2 mL 95% TFA solution in dichloromethane (2 × 1.5 h). HPLC purification afforded 3.2 mg heptamer **173** (3.5 µmol, 15% yield over 16 steps) and 4.8 mg hexamer **172** (6.1 µmol, 27% yield over 14 steps).

H-(Nsmd)$_7$-NH$_2$ (173): MALDI-TOF (matrix: DHB/CHCA 1:1, 0.1% TFA) m/z: 936 $[M+Na]^+$, 952 $[M+K]^+$.– Semipreparative HPLC (1% acetonitrile in water + 0.1% TFA over 10 min): t_R = 5.4 min, 85% purity.

H-(Nsmd)$_6$-NH$_2$ (172): MALDI-TOF (matrix: DHB/CHCA 1:1, 0.1% TFA) m/z: 808 $[M+Na]^+$, 824 $[M+K]^+$.– Semipreparative HPLC (1% acetonitrile in water + 0.1% TFA over 10 min): t_R = 4.5 min, 90% purity.

7. Abbreviations

α	optical rotation
Ac	acetyl
Ahx	6-aminohexanoic acid
Ar	aromatic
ATR	Attenuated Total Reflection
Bag-1	Bcl-2 associated athanogene-1
Boc	tert-butyloxycarbonyl
bs	broad singlet
Bu	butyl
°C	degrees Celsius
c	concentration
C_r	molar concentration per peptide bond
calc.	calculated
Cbz	carboxybenzyl
CD	Circular Dichroism
CHCA	α-Cyano-4-Hydroxy-Cinnamic Acid
cm	centimeter
conc.	concentrated
ctrl	control
CuAAc	Copper-catalyzed Alkyne-Azide Cycloaddition

δ	chemical shift
d	doublet (NMR) or cuvette length (CD)
DAD	Diode Array Detector
DCC	*N,N'*-dicyclohexylcarbodiimide
deg	degrees
DHB	2,5-dihydroxybenzoic acid
DIC	*N,N'*-diisopropylcarbodiimide
DIPEA	*N,N*-diisopropylethylamine
DKP	diketopiperazine
DMA	dimethylacetamide
DMEM	Dulbecco's Modified Eagle Medium
DMF	dimethylformamide
DMSO	dimethylsulfoxide
DNA	desoxyribonucleic acid
DTT	dithiothreitol
EI	electron impact ionization
Equiv.	equivalent
ER	Endoplasmic Reticulum
ESI	electrospray ionisation
Et	ethyl
e.g.	*exempli gratia* (for example)

et al.	*et alia* (and others)
etc.	*et cetera* (and so forth)
eV	electronvolt
FAB	Fast Atom Bombardement
FACS	Fluorescence Activated Cell Sorting
FITC	fluorescein isothiocyanate
FCS	Fetal Calf Serum
Fmoc	9*H*-fluoren-9-ylmethyloxycarbonyl
g	gram
GRP	Glucose-Regulated Protein
h	hour
HEK	Human Embryonic Kidney cells
HFIP	1,1,1,3,3,3-hexafluoroisopropanol
HGF	Hepatocyte Growth Factor
HIV	Human Immunodefficency Virus
HOBt	1-hydroxybenzotriazol
HPLC	High-Performance Liquid Chromatography
HRMS	High-Resolution Mass Spectrometry
HUVEC	Human Umbilical Vein Endothelial Cells
Hz	Hertz
IR	infrared

IRRAS	Infrared Reflection-Absorption Spectroscopy
J	coupling constant
J	Joule
L	liter
λ	wavelength
M	molar
μ	micro
m	multiplett (NMR), middle (IR), mili-
M	mega
MALDI-TOF	Matrix Assisted Laser Desorption/Ionisation – Time Of Flight
Me	methyl
mg	milligramm
min	minute
m.p	melting point
MRE	Mean Residue Ellipticity
MS	Mass Spectrometry
MW	microwave
n	number of assays, nano-
ν	frequency
NMR	Nuclear Magnetic Resonance spectroscopy
NMP	*N*-methyl-2-pyrrolidone

p	*para*
PBS	Phosphate-Buffer Saline
Ph	phenyl
ppm	parts per million
PS	polystyrene
q	quartet
quant.	quantitative
quin	quintuplet
r.t.	room temperature
s	singlet (NMR), strong (IR)
SCC	Squamous Cell Carcinoma
t	*tert*
t	triplet
θ	ellipticity
TCBC	2,4,6-trichlorobenzoyl chloride
TFA	trifluoroacetic acid
THF	tetrahydrofuran
TIS	triisopropylsilane
TMS	trimethylsilyl
t_R	retention time
Trt	trityl

UPR	Unfolded Protein Response
UV	ultraviolet
Vis	visible
vs	very strong (IR)
vw	very weak (IR)
w	weak (IR)

8. References

[1] R. M. J. Liskamp, D. T. S. Rijkers, J. A. W. Kruijtzer, J. Kemmink, *ChemBioChem* **2011**, *12*, 1626-1653. *Peptides and Proteins as a Continuing Exciting Source of Inspiration for Peptidomimetics.*

[2] A. Giannis, *Angew. Chem. Int. Ed.* **1993**, *32*, 1244-1267. *Peptidomimetics for Receptor Ligands Discovery, Development, and Medical Perspectives.*

[3] J. Gante, *Angew. Chem. Int. Ed.* **1994**, *33*, 1699-1720. *Peptidomimetics - Tailored Enzyme-Inhibitors.*

[4] J. Erickson, D. J. Neidhart, J. Vandrie, D. J. Kempf, X. C. Wang, D. W. Norbeck, J. J. Plattner, J. W. Rittenhouse, M. Turon, N. Wideburg, W. E. Kohlbrenner, R. Simmer, R. Helfrich, D. A. Paul, M. Knigge, *Science* **1990**, *249*, 527-533. *Design, Activity, and 2.8 a Crystal-Structure of a C2 Symmetrical Inhibitor Complexed to Hiv-1 Protease.*

[5] D. J. Kempf, K. C. Marsh, J. F. Denissen, E. Mcdonald, S. Vasavanonda, C. A. Flentge, B. E. Green, L. Fino, C. H. Park, X. P. Kong, N. E. Wideburg, A. Saldivar, L. Ruiz, W. M. Kati, H. L. Sham, T. Robins, K. D. Stewart, A. Hsu, J. J. Plattner, J. M. Leonard, D. W. Norbeck, *Proc. Natl. Acad. Sci. U. S. A.* **1995**, *92*, 2484-2488. *Abt-538 Is a Potent Inhibitor of Human-Immunodeficiency-Virus Protease and Has High Oral Bioavailability in Humans.*

[6] K. H. Hsieh, T. R. Lahann, R. C. Speth, *J. Med. Chem.* **1989**, *32*, 898-903. *Topographic Probes of Angiotensin and Receptor - Potent Angiotensin-Ii Agonist Containing Diphenylalanine and Long-Acting Antagonists Containing Biphenylalanine and 2 Indan Amino-Acid in Position 8.*

[7] M. H. V. Van Regenmortel, S. Muller, *Curr. Opin. Biotechnol.* **1998**, *9*, 377-382. *D-peptides as immunogens and diagnostic reagents.*

[8] C. Toniolo, M. Crisma, F. Formaggio, C. Peggion, *Biopolymers* **2001**, *60*, 396-419. *Control of peptide conformation by the Thorpe-Ingold effect (C(alpha)-tetrasubstitution).*

[9] D. Seebach, M. Overhand, F. N. M. Kuhnle, B. Martinoni, L. Oberer, U. Hommel, H. Widmer, *Helv. Chim. Acta* **1996**, *79*, 913-941. *beta-peptides: Synthesis by Arndt-Eistert homologation with concomitant peptide coupling. Structure determination by NMR and CD spectroscopy and by X-ray crystallography. Helical secondary structure of a beta-hexapeptide in solution and its stability towards pepsin.*

[10] D. Seebach, J. Gardiner, *Acc. Chem. Res.* **2008**, *41*, 1366-1375. *beta-Peptidic Peptidomimetics.*

[11] C. A. Olsen, *Biopolymers* **2011**, *96*, 561-566. *beta-Peptoid "Foldamers"-Why the Additional Methylene Unit?*

[12] Y. Gao, T. Kodadek, *Chem. Biol.* **2013**, *20*, 360-369. *Synthesis and Screening of Stereochemically Diverse Combinatorial Libraries of Peptide Tertiary Amides.*

[13] R. C. Elgersma, T. Meijneke, R. de Jong, A. J. Brouwer, G. Posthuma, D. T. S. Rijkers, R. M. J. Liskamp, *Org. Biomol. Chem.* **2006**, *4*, 3587-3597. *Synthesis and structural investigations of N-alkylated beta-peptidosulfonamide-peptide hybrids of*

the amyloidogenic amylin(20-29) sequence: implications of supramolecular folding for the design of peptide-based bionanomaterials.

[14] P. Claudon, A. Violette, K. Lamour, M. Decossas, S. Fournel, B. Heurtault, J. Godet, Y. Mely, B. Jamart-Gregoire, M. C. Averlant-Petit, J. P. Briand, G. Duportail, H. Monteil, G. Guichard, *Angew. Chem. Int. Ed.* **2010**, *49*, 333-336. *Consequences of Isostructural Main-Chain Modifications for the Design of Antimicrobial Foldamers: Helical Mimics of Host-Defense Peptides Based on a Heterogeneous Amide/Urea Backbone.*

[15] Z. H. Ke, H. F. Chow, M. C. Chan, Z. F. Liu, K. H. Sze, *Org. Lett.* **2012**, *14*, 394-397. *Head-to-Tail Dimerization and Organogelating Properties of Click Peptidomimetics.*

[16] D. Fürniß, Dissertation, Karlsruher Institut für Technologie (KIT) **2013**. *Neue molekulare Transporter auf Polyamin- und Peptoidbasis.*

[17] R. J. Simon, R. S. Kania, R. N. Zuckermann, V. D. Huebner, D. A. Jewell, S. Banville, S. Ng, L. Wang, S. Rosenberg, C. K. Marlowe, D. C. Spellmeyer, R. Y. Tan, A. D. Frankel, D. V. Santi, F. E. Cohen, P. A. Bartlett, *Proc. Natl. Acad. Sci. U. S. A.* **1992**, *89*, 9367-9371. *Peptoids - a Modular Approach to Drug Discovery.*

[18] P. S. Farmer, E. J. Ariens, *Trends Pharmacol. Sci.* **1982**, *3*, 362-365. *Speculations on the Design of Non-Peptidic Peptidomimetics.*

[19] S. M. Miller, R. J. Simon, S. Ng, R. N. Zuckermann, J. M. Kerr, W. H. Moos, *Bioorg. Med. Chem. Lett.* **1994**, *4*, 2657-2662. *Proteolytic Studies of Homologous Peptide and N-Substituted Glycine Peptoid Oligomers.*

[20] Y. U. Kwon, T. Kodadek, *J. Am. Chem. Soc.* **2007**, *129*, 1508-1509. *Quantitative evaluation of the relative cell permeability of peptoids and peptides.*

[21] R. N. Zuckermann, J. M. Kerr, S. B. H. Kent, W. H. Moos, *J. Am. Chem. Soc.* **1992**, *114*, 10646-10647. *Efficient Method for the Preparation of Peptoids [Oligo(N-Substituted Glycines)] by Submonomer Solid-Phase Synthesis.*

[22] R. B. Merrifield, *J. Am. Chem. Soc.* **1963**, *85*, 2149-2154. *Solid Phase Peptide Synthesis .1. Synthesis of a Tetrapeptide.*

[23] F. Rizzolo, C. Testa, D. Lambardi, M. Chorev, M. Chelli, P. Rovero, A. M. Papini, *J. Pept. Sci.* **2011**, *17*, 708-714. *Conventional and microwave-assisted SPPS approach: a comparative synthesis of PTHrP(1-34)NH(2).*

[24] R. Frank, *Tetrahedron* **1992**, *48*, 9217-9232. *Spot-Synthesis - an Easy Technique for the Positionally Addressable, Parallel Chemical Synthesis on a Membrane Support.*

[25] N. Heine, T. Ast, J. Schneider-Mergener, U. Reineke, L. Germeroth, H. Wenschuh, *Tetrahedron* **2003**, *59*, 9919-9930. *Synthesis and screening of peptoid arrays on cellulose membranes.*

[26] J. A. W. Kruijtzer, R. M. J. Liskamp, *Tetrahedron Lett.* **1995**, *36*, 6969-6972. *Synthesis in Solution of Peptoids Using Fmoc-Protected N-Substituted Glycines.*

[27] C. Caumes, T. Hjelmgaard, R. Remuson, S. Faure, C. Taillefumier, *Synthesis* **2011**, 257-264. *Highly Convenient Gram-Scale Solution-Phase Peptoid Synthesis and Orthogonal Side-Chain Post-Modification.*

[28] J. A. W. Kruijtzer, L. J. F. Hofmeyer, W. Heerma, C. Versluis, R. M. J. Liskamp, *Chem. Eur. J.* **1998**, *4*, 1570-1580. *Solid-phase syntheses of peptoids using Fmoc-*

protected N-substituted glycines: The synthesis of (retro) peptoids of Leu-enkephalin and substance P.

[29] M. A. Fara, J. J. Diaz-Mochon, M. Bradley, *Tetrahedron Lett.* **2006**, *47*, 1011-1014. *Microwave-assisted coupling with DIC/HOBt for the synthesis of difficult peptoids and fluorescently labelled peptides - a gentle heat goes a long way.*

[30] A. Unciti-Broceta, F. Diezmann, C. Y. Ou-Yang, M. A. Fara, M. Bradley, *Bioorg. Med. Chem.* **2009**, *17*, 959-966. *Synthesis, penetrability and intracellular targeting of fluorescein-tagged peptoids and peptide-peptoid hybrids.*

[31] A. S. Culf, R. J. Ouellette, *Molecules* **2010**, *15*, 5282-5335. *Solid-Phase Synthesis of N-Substituted Glycine Oligomers (alpha-Peptoids) and Derivatives.*

[32] G. M. Figliozzi, R. Goldsmith, S. C. Ng, S. C. Banville, R. N. Zuckermann, *Methods Enzymol.* **1996**, *267*, 437-447. *Synthesis of N-substituted glycine peptoid libraries.*

[33] H. J. Olivos, P. G. Alluri, M. M. Reddy, D. Salony, T. Kodadek, *Org. Lett.* **2002**, *4*, 4057-4059. *Microwave-assisted solid-phase synthesis of peptoids.*

[34] B. C. Gorske, S. A. Jewell, E. J. Guerard, H. E. Blackwell, *Org. Lett.* **2005**, *7*, 1521-1524. *Expedient synthesis and design strategies for new peptoid construction.*

[35] T. S. Burkoth, A. T. Fafarman, D. H. Charych, M. D. Connolly, R. N. Zuckermann, *J. Am. Chem. Soc.* **2003**, *125*, 8841-8845. *Incorporation of unprotected heterocyclic side chains into peptoid oligomers via solid-phase submonomer synthesis.*

[36] J. M. Holub, H. J. Jang, K. Kirshenbaum, *Org. Biomol. Chem.* **2006**, *4*, 1497-1502. *Clickity-click: highly functionalized peptoid oligomers generated by sequential conjugation reactions on solid-phase support.*

[37] D. Furniss, T. Mack, F. Hahn, S. B. L. Vollrath, K. Koroniak, U. Schepers, S. Bräse, *Beilstein J. Org. Chem.* **2013**, *9*, 56-63. *Peptoids and polyamines going sweet: Modular synthesis of glycosylated peptoids and polyamines using click chemistry.*

[38] S. A. Fowler, H. E. Blackwell, *Org. Biomol. Chem.* **2009**, *7*, 1508-1524. *Structure-function relationships in peptoids: Recent advances toward deciphering the structural requirements for biological function.*

[39] P. Armand, K. Kirshenbaum, A. Falicov, R. L. Dunbrack, K. A. Dill, R. N. Zuckermann, F. E. Cohen, *Fold. Des.* **1997**, *2*, 369-375. *Chiral N-substituted glycines can form stable helical conformations.*

[40] K. Kirshenbaum, A. E. Barron, R. A. Goldsmith, P. Armand, E. K. Bradley, K. T. V. Truong, K. A. Dill, F. E. Cohen, R. N. Zuckermann, *Proc. Natl. Acad. Sci. U. S. A.* **1998**, *95*, 4303-4308. *Sequence-specific polypeptoids: A diverse family of heteropolymers with stable secondary structure.*

[41] P. Armand, K. Kirshenbaum, R. A. Goldsmith, S. Farr-Jones, A. E. Barron, K. T. V. Truong, K. A. Dill, D. F. Mierke, F. E. Cohen, R. N. Zuckermann, E. K. Bradley, *Proc. Natl. Acad. Sci. U. S. A.* **1998**, *95*, 4309-4314. *NMR determination of the major solution conformation of a peptoid pentamer with chiral side chains.*

[42] C. W. Wu, K. Kirshenbaum, T. J. Sanborn, J. A. Patch, K. Huang, K. A. Dill, R. N. Zuckermann, A. E. Barron, *J. Am. Chem. Soc.* **2003**, *125*, 13525-13530. *Structural and spectroscopic studies of peptoid oligomers with alpha-chiral aliphatic side chains.*

[43] C. W. Wu, T. J. Sanborn, K. Huang, R. N. Zuckermann, A. E. Barron, *J. Am. Chem. Soc.* **2001**, *123*, 6778-6784. *Peptoid oligomers with alpha-chiral, aromatic side chains: Sequence requirements for the formation of stable peptoid helices.*

[44] J. R. Stringer, J. A. Crapster, I. A. Guzei, H. E. Blackwell, *J. Am. Chem. Soc.* **2011**, *133*, 15559-15567. *Extraordinarily Robust Polyproline Type I Peptoid Helices Generated via the Incorporation of alpha-Chiral Aromatic N-1-Naphthylethyl Side Chains.*

[45] K. Huang, C. W. Wu, T. J. Sanborn, J. A. Patch, K. Kirshenbaum, R. N. Zuckermann, A. E. Barron, I. Radhakrishnan, *J. Am. Chem. Soc.* **2006**, *128*, 1733-1738. *A threaded loop conformation adopted by a family of peptoid nonamers.*

[46] B. C. Gorske, H. E. Blackwell, *J. Am. Chem. Soc.* **2006**, *128*, 14378-14387. *Tuning peptoid secondary structure with pentafluoroaromatic functionality: A new design paradigm for the construction of discretely folded peptoid structures.*

[47] B. C. Gorske, J. R. Stringer, B. L. Bastian, S. A. Fowler, H. E. Blackwell, *J. Am. Chem. Soc.* **2009**, *131*, 16555-16567. *New Strategies for the Design of Folded Peptoids Revealed by a Survey of Noncovalent Interactions in Model Systems.*

[48] N. H. Shah, G. L. Butterfoss, K. Nguyen, B. Yoo, R. Bonneau, D. L. Rabenstein, K. Kirshenbaum, *J. Am. Chem. Soc.* **2008**, *130*, 16622-16632. *Oligo(N-aryl glycines): A New Twist on Structured Peptoids.*

[49] C. Caumes, O. Roy, S. Faure, C. Taillefumier, *J. Am. Chem. Soc.* **2012**, *134*, 9553-9556. *The Click Triazolium Peptoid Side Chain: A Strong cis-Amide Inducer Enabling Chemical Diversity.*

[50] J. A. Crapster, I. A. Guzei, H. E. Blackwell, *Angew. Chem. Int. Ed.* **2013**, *52*, 5079-5084. *A Peptoid Ribbon Secondary Structure.*

[51] R. M. Kohli, C. T. Walsh, M. D. Burkart, *Nature* **2002**, *418*, 658-661. *Biomimetic synthesis and optimization of cyclic peptide antibiotics.*

[52] H. Kessler, B. Diefenbach, D. Finsinger, A. Geyer, M. Gurrath, S. L. Goodman, G. Holzemann, R. Haubner, A. Jonczyk, G. Muller, E. G. vonRoedern, J. Wermuth, *Lett. Pept. Sci.* **1995**, *2*, 155-160. *Design of superactive and selective integrin receptor antagonists containing the RGD sequence.*

[53] Z. D. Shi, K. Lee, C. Q. Wei, L. R. Roberts, K. M. Worthy, R. J. Fisher, T. R. Burke, *J. Med. Chem.* **2004**, *47*, 788-791. *Synthesis of a 5-methylindolyl-containing macrocycle that displays ultrapotent Grb2 SH2 domain-binding affinity.*

[54] S. B. Y. Shin, B. Yoo, L. J. Todaro, K. Kirshenbaum, *J Am Chem Soc* **2007**, *129*, 3218-3225. *Cyclic peptoids.*

[55] M. L. Huang, S. B. Y. Shin, M. A. Benson, V. J. Torres, K. Kirshenbaum, *ChemMedChem* **2012**, *7*, 114-122. *A Comparison of Linear and Cyclic Peptoid Oligomers as Potent Antimicrobial Agents.*

[56] S. B. L. Vollrath, C. H. Hu, S. Bräse, K. Kirshenbaum, *Chem. Commun.* **2013**, *49*, 2317-2319. *Peptoid nanotubes: an oligomer macrocycle that reversibly sequesters water via single-crystal-to-single-crystal transformations.*

[57] I. Izzo, G. Ianniello, C. De Cola, B. Nardone, L. Erra, G. Vaughan, C. Tedesco, F. De Riccardis, *Org. Lett.* **2013**, *15*, 598-601. *Structural Effects of Proline Substitution and Metal Binding on Hexameric Cyclic Peptoids.*

[58] B. Vaz, L. Brunsveld, *Org. Biomol. Chem.* **2008**, *6*, 2988-2994. *Stable helical peptoids via covalent side chain to side chain cyclization.*

[59] J. M. Holub, H. J. Jang, K. Kirshenbaum, *Org. Lett.* **2007**, *9*, 3275-3278. *Fit to be tied: Conformation-directed macrocyclization of peptoid foldamers.*

[60] S. N. Khan, A. Kim, R. H. Grubbs, Y. U. Kwon, *Org. Lett.* **2011**, *13*, 1582-1585. *Ring-Closing Metathesis Approaches for the Solid-Phase Synthesis of Cyclic Peptoids.*

[61] S. Chirayil, K. J. Luebke, *Tetrahedron Lett.* **2012**, *53*, 726-729. *Cyclization of peptoids by formation of boronate esters.*

[62] L. Guo, D. H. Zhang, *J. Am. Chem. Soc.* **2009**, *131*, 18072-+. *Cyclic Poly(alpha-peptoid)s and Their Block Copolymers from N-Heterocyclic Carbene-Mediated Ring-Opening Polymerizations of N-Substituted N-Carboxylanhydrides.*

[63] S. B. L. Vollrath, S. Bräse, K. Kirshenbaum, *Chem. Sci.* **2012**, *3*, 2726-2731. *Twice tied tight: Enforcing conformational order in bicyclic peptoid oligomers.*

[64] J. Sun, R. N. Zuckermann, *ACS Nano* **2013**, *7*, 4715-4732. *Peptoid Polymers: A Highly Designable Bioinspired Material.*

[65] A. M. Rosales, H. K. Murnen, R. N. Zuckermann, R. A. Segalman, *Macromolecules* **2010**, *43*, 5627-5636. *Control of Crystallization and Melting Behavior in Sequence Specific Polypeptoids.*

[66] W. van Zoelen, R. N. Zuckermann, R. A. Segalman, *Macromolecules* **2012**, *45*, 7072-7082. *Tunable Surface Properties from Sequence-Specific Polypeptoid-Polystyrene Block Copolymer Thin Films.*

[67] K. T. Nam, S. A. Shelby, P. H. Choi, A. B. Marciel, R. Chen, L. Tan, T. K. Chu, R. A. Mesch, B. C. Lee, M. D. Connolly, C. Kisielowski, R. N. Zuckermann, *Nature Materials* **2010**, *9*, 454-460. *Free-floating ultrathin two-dimensional crystals from sequence-specific peptoid polymers.*

[68] B. Sanii, R. Kudirka, A. Cho, N. Venkateswaran, G. K. Olivier, A. M. Olson, H. Tran, R. M. Harada, L. Tan, R. N. Zuckermann, *J. Am. Chem. Soc.* **2011**, *133*, 20808-20815. *Shaken, Not Stirred: Collapsing a Peptoid Monolayer To Produce Free-Floating, Stable Nanosheets.*

[69] R. Kudirka, H. Tran, B. Sanii, K. T. Nam, P. H. Choi, N. Venkateswaran, R. Chen, S. Whitelam, R. N. Zuckermann, *Biopolymers* **2011**, *96*, 586-595. *Folding of a Single-Chain, Information-Rich Polypeptoid Sequence into a Highly Ordered Nanosheet.*

[70] H. K. Murnen, A. M. Rosales, J. N. Jaworsk, R. A. Segalman, R. N. Zuckermann, *J. Am. Chem. Soc.* **2010**, *132*, 16112-16119. *Hierarchical Self-Assembly of a Biomimetic Diblock Copolypeptoid into Homochiral Superhelices.*

[71] M. Goodman, M. Bhumralkar, E. A. Jefferson, J. Kwak, E. Locardi, *Biopolymers* **1998**, *47*, 127-142. *Collagen mimetics.*

[72] A. R. Statz, R. J. Meagher, A. E. Barron, P. B. Messersmith, *J. Am. Chem. Soc.* **2005**, *127*, 7972-7973. *New peptidomimetic polymers for antifouling surfaces.*

[73] C. L. Chen, J. H. Qi, R. N. Zuckermann, J. J. DeYoreo, *J. Am. Chem. Soc.* **2011**, *133*, 5214-5217. *Engineered Biomimetic Polymers as Tunable Agents for Controlling CaCO3 Mineralization.*

[74] G. Maayan, M. D. Ward, K. Kirshenbaum, *Proc. Natl. Acad. Sci. U. S. A.* **2009**, *106*, 13679-13684. *Folded biomimetic oligomers for enantioselective catalysis.*

[75] W. L. Zhu, Y. M. Song, Y. Park, K. H. Park, S. T. Yang, J. I. Kim, I. S. Park, K. S. Hahm, S. Y. Shin, *Biochim. Biophys. Acta-Biomembranes* **2007**, *1768*, 1506-1517. *Substitution of the leucine zipper sequence in melittin with peptoid residues affects self-association, cell selectivity, and mode of action.*

[76] F. S. Nandel, A. Saini, *Macromol. Theory Simul.* **2007**, *16*, 295-303. *Conformational study of short peptoid models for future applications as potent antimicrobial compounds.*

[77] R. Kapoor, M. W. Wadman, M. T. Dohm, A. M. Czyzewski, A. M. Spormann, A. E. Barron, *Antimicrob. Agents Chemother.* **2011**, *55*, 3054-3057. *Antimicrobial Peptoids Are Effective against Pseudomonas aeruginosa Biofilms.*

[78] N. P. Chongsiriwatana, J. A. Patch, A. M. Czyzewski, M. T. Dohm, A. Ivankin, D. Gidalevitz, R. N. Zuckermann, A. E. Barron, *P Natl Acad Sci USA* **2008**, *105*, 2794-2799. *Peptoids that mimic the structure, function, and mechanism of helical antimicrobial peptides.*

[79] D. Comegna, M. Benincasa, R. Gennaro, I. Izzo, F. De Riccardis, *Bioorgan Med Chem* **2010**, *18*, 2010-2018. *Design, synthesis and antimicrobial properties of non-hemolytic cationic alpha-cyclopeptoids.*

[80] S. L. Seurynck-Servoss, M. T. Dohm, A. E. Barron, *Biochemistry* **2006**, *45*, 11809-11818. *Effects of including an N-terminal insertion region and arginine-mimetic side chains in helical peptoid analogues of lung surfactant protein B.*

[81] N. J. Brown, C. W. Wu, S. L. Seurynck-Servoss, A. E. Barron, *Biochemistry* **2008**, *47*, 1808-1818. *Effects of hydrophobic helix length and side chain chemistry on biomimicry in peptoid analogues of SP-C.*

[82] R. C. Elgersma, G. E. Mulder, J. A. W. Kruijtzer, G. Posthuma, D. T. S. Rijkers, R. M. J. Liskamp, *Bioorg. Med. Chem. Lett.* **2007**, *17*, 1837-1842. *Transformation of the amyloidogenic peptide amylin(20-29) into its corresponding peptoid and retropeptoid: Access to both an amyloid inhibitor and template for self-assembled supramolecular tapes.*

[83] P. G. Alluri, M. M. Reddy, K. Bachhawat-Sikder, H. J. Olivos, T. Kodadek, *J Am Chem Soc* **2003**, *125*, 13995-14004. *Isolation of protein ligands from large peptoid libraries.*

[84] P. Alluri, B. Liu, P. Yu, X. S. Xiao, T. Kodadek, *Mol. Biosyst.* **2006**, *2*, 568-579. *Isolation and characterization of coactivator-binding peptoids from a combinatorial library.*

[85] C. Garcia-Martinez, M. Humet, R. Planells-Cases, A. Gomis, M. Caprini, F. Viana, E. De la Pena, F. Sanchez-Baeza, T. Carbonell, C. De Felipe, E. Perez-Paya, C. Belmonte, A. Messeguer, A. Ferrer-Montiel, *Proc. Natl. Acad. Sci. U. S. A.* **2002**, *99*, 2374-2379. *Attenuation of thermal nociception and hyperalgesia by VR1 blockers.*

[86] R. N. Zuckermann, E. J. Martin, D. C. Spellmeyer, G. B. Stauber, K. R. Shoemaker, J. M. Kerr, G. M. Figliozzi, D. A. Goff, M. A. Siani, R. J. Simon, S. C. Banville, E. G. Brown, L. Wang, L. S. Richter, W. H. Moos, *J. Med. Chem.* **1994**, *37*, 2678-2685. *Discovery of Nanomolar Ligands for 7-Transmembrane G-Protein-Coupled Receptors from a Diverse N-(Substituted)Glycine Peptoid Library.*

[87] M. M. Reddy, R. Wilson, J. Wilson, S. Connell, A. Gocke, L. Hynan, D. German, T. Kodadek, *Cell* **2011**, *144*, 132-142. *Identification of Candidate IgG Biomarkers for Alzheimer's Disease via Combinatorial Library Screening.*

[88] A. Y. Yam, X. M. Wang, C. M. Gao, M. D. Connolly, R. N. Zuckermann, T. Bleu, J. Hall, J. P. Fedynyshyn, S. Allauzen, D. Peretz, C. M. Salisbury, *Biochemistry* **2011**, *50*, 4322-4329. *A Universal Method for Detection of Amyloidogenic Misfolded Proteins.*

[89] T. Hara, S. R. Durell, M. C. Myers, D. H. Appella, *J Am Chem Soc* **2006**, *128*, 1995-2004. *Probing the structural requirements of peptoids that inhibit HDM2-p53 interactions.*

[90] J. C. Hooks, J. P. Matharage, D. G. Udugamasooriya, *Biopolymers* **2011**, *96*, 567-577. *Development of Homomultimers and Heteromultimers of Lung Cancer-Specific Peptoids.*

[91] P. M. Levine, K. Imberg, M. J. Garabedian, K. Kirshenbaum, *J. Am. Chem. Soc.* **2012**, *134*, 6912-6915. *Multivalent Peptidomimetic Conjugates: A Versatile Platform for Modulating Androgen Receptor Activity.*

[92] R. Brasseur, G. Divita, *Biochim. Biophys. Acta-Biomembranes* **2010**, *1798*, 2177-2181. *Happy birthday cell penetrating peptides: Already 20 years.*

[93] A. D. Frankel, C. O. Pabo, *Cell* **1988**, *55*, 1189-1193. *Cellular Uptake of the Tat Protein from Human Immunodeficiency Virus.*

[94] J. E. Murphy, T. Uno, J. D. Hamer, F. E. Cohen, V. Dwarki, R. N. Zuckermann, *Proc. Natl. Acad. Sci. U. S. A.* **1998**, *95*, 1517-1522. *A combinatorial approach to the discovery of efficient cationic peptoid reagents for gene delivery.*

[95] C. Y. Huang, T. Uno, J. E. Murphy, S. Lee, J. D. Hamer, J. A. Escobedo, F. E. Cohen, R. Radhakrishnan, V. Dwarki, R. N. Zuckermann, *Chem Biol* **1998**, *5*, 345-354. *Lipitoids - novel cationic lipids for cellular delivery of plasmid DNA in vitro.*

[96] P. A. Wender, D. J. Mitchell, K. Pattabiraman, E. T. Pelkey, L. Steinman, J. B. Rothbard, *Proc. Natl. Acad. Sci. U. S. A.* **2000**, *97*, 13003-13008. *The design, synthesis, and evaluation of molecules that enable or enhance cellular uptake: Peptoid molecular transporters.*

[97] T. Schroder, N. Niemeier, S. Afonin, A. S. Ulrich, H. F. Krug, S. Bräse, *J. Med. Chem.* **2008**, *51*, 376-379. *Peptoidic amino- and guanidinium-carrier systems: Targeted drug delivery into the cell cytosol or the nucleus.*

[98] U. Sternberg, E. Birtalan, I. Jakovkin, B. Luy, U. Schepers, S. Bräse, C. Muhle-Goll, *Org. Biomol. Chem.* **2013**, *11*, 640-647. *Structural characterization of a peptoid with lysine-like side chains and biological activity using NMR and computational methods.*

[99] K. Eggenberger, E. Birtalan, T. Schroder, S. Bräse, P. Nick, *ChemBioChem* **2009**, *10*, 2504-2512. *Passage of Trojan Peptoids into Plant Cells.*

[100] D. K. F. Kölmel, D.; Susanto, S.; Lauer, A.; Grabher, C.; Bräse, S.; Schepers, U., *Pharmaceuticals* **2012**, *5*, 1265-1281. *Cell Penetrating Peptoids (CPPos): Synthesis of a Small Combinatorial Library by Using IRORI MiniKans*

[101] D. T. Thielemann, A. T. Wagner, E. Rosch, D. K. Kölmel, J. G. Heck, B. Rudat, M. Neumaier, C. Feldmann, U. Schepers, S. Bräse, P. W. Roesky, *J. Am. Chem. Soc.*

2013, *135*, 7454-7457. *Luminescent Cell-Penetrating Pentadecanuclear Lanthanide Clusters.*

[102] B. C. Lee, R. N. Zuckermann, K. A. Dill, *J. Am. Chem. Soc.* **2005**, *127*, 10999-11009. *Folding a nonbiological polymer into a compact multihelical structure.*

[103] B. C. Lee, T. K. Chu, K. A. Dill, R. N. Zuckermann, *J. Am. Chem. Soc.* **2008**, *130*, 8847-8855. *Biomimetic nanostructures: Creating a high-affinity zinc-binding site in a folded nonbiological polymer.*

[104] C. Birchmeier, W. Birchmeier, E. Gherardi, G. F. Vande Woude, *Nat. Rev. Mol. Cell Bio.* **2003**, *4*, 915-925. *Met, metastasis, motility and more.*

[105] V. Orian-Rousseau, L. F. Chen, J. P. Sleeman, P. Herrlich, H. Ponta, *Genes Dev.* **2002**, *16*, 3074-3086. *CD44 is required for two consecutive steps in HGF/c-Met signaling.*

[106] A. Matzke, P. Herrlich, H. Ponta, W. Orian-Rousseau, *Cancer Res.* **2005**, *65*, 6105-6110. *A five-amino-acid peptide blocks Met- and Ron-dependent cell migration.*

[107] P. A. Townsend, A. Stephanou, G. Packham, D. S. Latchman, *Int. J. Biochem. Cell Biol.* **2005**, *37*, 251-259. *BAG-1: a multi-functional pro-survival molecule.*

[108] Y. Nadler, R. L. Camp, J. M. Giltnane, C. Moeder, D. L. Rimm, H. M. Kluger, Y. Kluger, *Breast Cancer Res.* **2008**, *10. Expression patterns and prognostic value of Bag-1 and Bcl-2 in breast cancer.*

[109] D. Maddalo, A. Neeb, K. Jehle, K. Schmitz, C. Muhle-Goll, L. Shatkina, T. V. Walther, A. Bruchmann, S. M. Gopal, W. Wenzel, A. S. Ulrich, A. C. B. Cato, *PLOS One* **2012**, *7. A Peptidic Unconjugated GRP78/BiP Ligand Modulates the Unfolded Protein Response and Induces Prostate Cancer Cell Death.*

[110] R. K. Reddy, C. H. Mao, P. Baumeister, R. C. Austin, R. J. Kaufman, A. S. Lee, *J. Biol. Chem.* **2003**, *278*, 20915-20924. *Endoplasmic reticulum chaperone protein GRP78 protects cells from apoptosis induced by topoisomerase inhibitors - Role of ATP binding site in suppression of caspase-7 activation.*

[111] A. Kardosh, E. B. Golden, P. Pyrko, J. Uddin, F. M. Hofman, T. C. Chen, S. G. Louie, N. A. Petasis, A. H. Schonthal, *Cancer Res.* **2008**, *68*, 843-851. *Aggravated endoplasmic reticulum stress as a basis for enhanced glioblastoma cell killing by bortezomib in combination with celecoxib or its non-coxib analogue, 2,5-dimethyl-celecoxib.*

[112] Y. Fu, A. S. Lee, *Cancer Biol. Ther.* **2006**, *5*, 741-744. *Glucose regulated proteins in cancer progression, drug resistance and immunotherapy.*

[113] M. Giraud, F. Cavelier, J. Martinez, *J. Pept. Sci.* **1999**, *5*, 457-461. *A side-reaction in the SPPS of Trp-containing peptides.*

[114] T. Uno, E. Beausoleil, R. A. Goldsmith, B. H. Levine, R. N. Zuckermann, *Tetrahedron Lett.* **1999**, *40*, 1475-1478. *New submonomers for poly N-substituted glycines (peptoids).*

[115] M. S. Bjelakovic, D. M. Godjevac, D. R. Milic, *Carbon* **2007**, *45*, 2260-2265. *Synthesis and antioxidant properties of fullero-steroidal covalent conjugates.*

[116] R. Gutierrez-Abad, O. Illa, R. M. Ortuno, *Org. Lett.* **2010**, *12*, 3148-3151. *Synthesis of Chiral Cyclobutane Containing C-3-Symmetric Peptide Dendrimers.*

[117] E. Nnanabu, K. Burgess, *Org. Lett.* **2006**, *8*, 1259-1262. *Cyclic semipeptoids: Peptoid-organic hybrid macrocycles.*

[118] R. Castonguay, C. Lherbet, J. W. Keillor, *Bioorg. Med. Chem.* **2002**, *10*, 4185-4191. *Mapping of the active site of rat kidney gamma-glutamyl transpeptidase using activated esters and their amide derivatives.*

[119] S. B. L. Vollrath, Dissertation, Karlsruher Institut für Technologie (KIT) **2012**. *Synthese Funktioneller Peptoide und Peptoid-Macrocyclen.*

[120] P. Sieber, B. Riniker, *Tetrahedron Lett.* **1991**, *32*, 739-742. *Protection of Carboxamide Functions by the Trityl Residue - Application to Peptide-Synthesis.*

[121] C. Cardenal, S. B. L. Vollrath, U. Schepers, S. Bräse, *Helv. Chim. Acta* **2012**, *95*, 2237-2248. *Synthesis of Functionalized Glutamine- and Asparagine-Type Peptoids - Scope and Limitations.*

[122] K. K. Hansen, B. Grosch, S. Greiveldinger-Poenaru, P. A. Bartlett, *J. Org. Chem.* **2003**, *68*, 8465-8470. *Synthesis and evaluation of macrocyclic transition state analogue inhibitors for alpha-chymotrypsin.*

[123] E. Birtalan, Dissertation, Karlsruher Institut für Technologie (KIT) **2009**. *Chemische Biologie von Molekularen Transportern auf Peptoid-Basis.*

[124] M. Park, M. Wetzler, T. S. Jardetzky, A. E. Barron, *PLOS One* **2013**, *8*. *A Readily Applicable Strategy to Convert Peptides to Peptoid-based Therapeutics.*

[125] P. S. Steeg, *Nat. Med.* **2006**, *12*, 895-904. *Tumor metastasis: mechanistic insights and clinical challenges.*

[126] R. W. Riddoch, P. Schaffer, J. F. Valliant, *Bioconjug. Chem.* **2006**, *17*, 226-235. *A solid-phase labeling strategy for the preparation of technetium and rhenium bifunctional chelate complexes and associated peptide conjugates.*

[127] P. Edman, *Acta Chem. Scand.* **1950**, *4*, 283-293. *Method for Determination of the Amino Acid Sequence in Peptides.*

[128] C. Mas-Moruno, L. J. Cruz, P. Mora, A. Francesch, A. Messeguer, E. Perez-Paya, F. Albericio, *J. Med. Chem.* **2007**, *50*, 2443-2449. *Smallest peptoids with antiproliferative activity on human neoplastic cells.*

[129] N. J. Greenfield, *Nat. Protoc.* **2006**, *1*, 2876-2890. *Using circular dichroism spectra to estimate protein secondary structure.*

[130] S. Beychok, *Science* **1966**, *154*, 1288-1299. *Circular Dichroism of Biological Macromolecules.*

[131] T. J. Sanborn, C. W. Wu, R. N. Zuckerman, A. E. Barron, *Biopolymers* **2002**, *63*, 12-20. *Extreme stability of helices formed by water-soluble poly-N-substituted glycines (polypeptoids) with alpha-chiral side chains.*

[132] D. Kölmel, Dissertation, Karlsruher Institut für Technologie (KIT) **2013**. *Chemische Biologie von neuen zellgängigen Peptoiden und Synthese fluoriger Farbstoffe.*

[133] Autorenkollektiv, *Organikum*, D. V. der Wissentschaften, Berlin, **1990**.

[134] T. Vojkovsky, *Pept. Res.* **1995**, *8*, 236-237. *Detection of secondary amines on solid phase.*

[135] W. C. Still, M. Kahn, A. Mitra, *J. Org. Chem.* **1978**, *43*, 2923-2925. *Rapid Chromatographic Technique for Preparative Separations with Moderate Resolution.*

[136] M. J. Dixon, A. Nathubhai, O. A. Andersen, D. M. van Aalten, I. M. Eggleston, *Org. Biomol. Chem.* **2009**, *7*, 259-268. *Solid-phase synthesis of cyclic peptide chitinase inhibitors: SAR of the argifin scaffold.*

[137] J. D. White, Y. Li, D. C. Ihle, *J. Org. Chem.* **2010**, *75*, 3569-3577. *Tandem Intramolecular Photocycloaddition−Retro-Mannich Fragmentation as a Route to Spiro[pyrrolidine-3,3'-oxindoles]. Total Synthesis of (±)-Coerulescine, (±)-Horsfiline, (±)-Elacomine, and (±)-6-Deoxyelacomine.*

[138] D. Castagnolo, F. Raffi, G. Giorgi, M. Botta, *Eur. J. Org. Chem.* **2009**, 334-337. *Macrocyclization of Di-Boc-guanidino-alkylamines Related to Guazatine Components: Discovery and Synthesis of Innovative Macrocyclic Amidinoureas.*

[139] M. Weitman, K. Lerman, A. Nudelman, D. T. Major, A. Hizi, A. Herschhorn, *Eur. J. Med. Chem.* **2011**, *46*, 447-467. *Structure activity relationship studies of 1-(4-chloro-2,5-dimethoxyphenyl)-3-(3-propoxypropyl)thiourea, a non-nucleoside reverse transcriptase inhibitor of human immunodeficiency virus type-1.*

[140] O. Josse, D. Labar, J. Marchand-Brynaert, *Synthesis* **1999**, 404-406. *A convenient synthesis of ethyl 3-aminopropanedithioate (beta-alanine ethyl dithioester).*

[141] T. Schroder, K. Schmitz, N. Niemeier, T. S. Balaban, H. F. Krug, U. Schepers, S. Bräse, *Bioconjug. Chem.* **2007**, *18*, 342-354. *Solid-phase synthesis, bioconjugation, and toxicology of novel cationic oligopeptoids for cellular drug delivery.*

[142] R. A. Gardner, G. Ghobrial, S. A. Naser, O. Phanstiel, *J. Med. Chem.* **2004**, *47*, 4933-4940. *Synthesis and biological evaluation of new acinetoferrin homologues for use as iron transport probes in mycobacteria.*

9. Appendix

9.1 Lebenslauf

Persönliche Daten

Name: Carmen Cardenal Pac
Adresse: Kaiserallee 105, 76185 Karlsruhe
E-Mail: ccardenalp@gmail.com

Geburtsdatum: 07.12.1986
Geburtsort: Barcelona (Spanien)

Schulische Ausbildung und akademischer Werdegang

10/2010 – 12/2013 **Karlsruher Institut für Technologie** (Karlsruhe, Deutschland)
Promotion im Arbeitskreis von Prof. Dr. Stefan Bräse am Institut für Organische Chemie und Mitglied der „Biointerfaces International Graduate School (BIF-IGS)"
Thema: *Design, Synthesis and Evaluation of Highly Functionalized Peptoids as Antitumor Peptidomimetics.*

09/2009 – 08/2010 **Bayer CropScience AG** (Frankfurt am Main, Deutschland)
Studiumsabschlussarbeit.
Thema: *Synthesis of Bioactive sesquiterpene analogs* via *oganometalic catalysis.*

10/2004 – 06/2009 **Institut Químic de Sarrià (IQS), Universitat Ramón Llull** (Barcelona, Spanien)
Studium der Chemie. Vertiefung in Organische Chemie.
Studienabschluss: Licenciatura (entspricht im deutschen System dem „Diplom")

09/1998 – 06/2004 **Joan Boscà Gymnasium** (Barcelona, Spanien)
„Bachillerato/PAU" (entspricht im deutschen System dem „Abitur"). Schwerpunkt auf Naturwissenschaften.

Stipendien

09/2011 – 12/2013 Promotionsstipendium des **La Caixa-Deutschen Akademischen Austauschdienstes (DAAD)**

09/2009 – 08/2010 ERASMUS-Praktikum, Stipendium der **„Generalitat de Catalunya"** für einen Forschungsaufenthalt in Deutschland

Sprachen

Spanisch:	Muttersprache
Katalanisch:	Muttersprache
Englisch:	Verhandlungssicher; *Certificate of Proficiency in English from Cambridge University* (C2) (2007)
Französisch:	Fließend, *Diplôme du DELF B2* (2009)
Deutsch:	Fließend

9.2 Publications and Conferences

Publications

C. Cardenal, S. B. L. Vollrath, U. Schepers, S. Bräse, *Helv. Chim. Acta* **2012**, *95*, 2237-2248. *Synthesis of Functionalized Glutamine- and Asparagine-Type Peptoids – Scope and Limitations.*

Conference Contributions

ORAL PRESENTATIONS

C. Cardenal, BioInterfaces International Graduate School - Summer Retreat 2013, Gültstein, Germany, 02.09 – 04.09.2013. *Synthesis and Evaluation of Peptoids as antitumor peptidomimetics.*

POSTER PRESENTATIONS

C. Cardenal, S. Bräse, *BioInterfaces International Graduate School- Summer Retreat 2011*, Bad Herrenalb, Germany, 31.08 – 02.09.2011. *Peptoids as peptidomimetics.*

C. Cardenal, S. Bräse, *8th Peptoid Summit,* Berkeley, USA, 09.08-10.08.2012. *Peptoids as antitumor peptidomimetics.*

C. Cardenal, S. Bräse, BioInterfaces International Graduate School- Summer Retreat 2012, Gültstein, Germany, 03.09 – 05.09.2012. *Peptoids as antitumor peptidomimetics.*

9.3 Aknowledgements

Zuallererst möchte ich mich bei meinem Doktorvater Prof. Stefan Bräse für die freundliche Aufnahme in seine Arbeitsgruppe sowie die interessante Themenstellung ganz herzlich bedanken. Darüber hinaus bedanke ich mich für seine fachliche Betreuung sowie die große Freiheit und Unterstützung bei der Umsetzung meiner eigenen Ideen.

Prof. Wagenknecht danke ich für die freundliche Übernahme des Korreferats.

Bei den Mitgliedern meines TAC Komitees: PD Dr. Ute Schepers, PD Dr. Veronique Orian-Rousseau und Prof. Katja Schmitz bedanke ich mich auch ganz herzlich für ihre Betreuung und die hilfreichen Diskussionen.

Ein großes Dankeschön gebührt Frau Christiane Lampert und Frau Désirée Witt für ihre Hilfsbereitschaft bei allen organisatorischen und bürokratischen Fragen.

Für die finanzielle Unterstützung dieser Arbeit danke ich La Caixa und dem Deutschen Akademischen Austauschdienst (DAAD).

Bei allen Mitarbeitern des Instituts die zu dieser Arbeit beigetragen haben möchte ich mich auch ganz herzlich bedanken, insbesondere bei Frau Pia Lang, Frau Tanja Ohmert, Frau Angelika Kernert und Frau Ingrid Roßnagel aus der Analytikabteilung. Richard von Budberg und Jakob Meyer danke ich für die Reparaturen von Glas- und Laborgeräten sowie Dominic Lüthjohann für seine schnelle Antwort bei der Lösung von Computerproblemen.

Ein besonderes Dankeschön gilt Karolin Niessner für ihre riesige Hilfe bei HPLC Reinigungen und ihre Geduld.

Bei Dr. Anja von Au, Mark Schmitt und Dr. Antje Neeb bedanke ich mich ganz herzlich für die Durchführung der biologischen Tests und die bereichernde kooperative Arbeit. Sie waren ein essentieller Teil dieses Projekts.

Für die kritische Korrektur dieses Manuskripts danke ich Martina Austeri, Dominik Kölmel und Katharina Peschko.

Meinen Vertiefungsstudentinnen Katharina Peschko und Andrea Lauer danke ich ebenfalls für ihren Beitrag zu dieser Dissertation.

An dieser Stelle möchte ich mich auch bei der ganzen Arbeitsgruppe Bräse für die freundliche Arbeitsatmosphäre und die wunderschöne Zeit innerhalb und außerhalb des Labors bedanken. Ein besondere Dank gebührt dem „Peptoid Team" Sidonie Vollrath, Dominik Kölmel, Daniel Fürniß und Katharina Peschko für die gute Zusammenarbeit und die hilfreichen Tipps und Diskussionen. Außerdem danke ich meinen Laborkollegen Anne Meister und Dominik Kölmel für ihre Hilfsbereitschaft und die schöne Zeit im Labor. Weiterhin danke ich Andreas Hafner und Joshua Kramer für die schöne Atmosphäre und moralische Unterstützung bei jedem Schritt und Atemzug des Schreibprozesses.

Ein riesen Dankeschön an meine Mitbewohner Daniel, Daniela, Stefan, Frank, Klaus, Rahel, Olga und Andres, weil sie meine kleine Familie in Deutschland während dieser drei Jahre waren.

Finalment, el major agraïment és per als meus pares, que m'han donat totes les possibilitats de formació que he desitjat i més. Gràcies pel vostre suport i la vostra confiança. Al meu germà Felip li agraeixo el seu bon humor i suport incondicionals. I a la resta de familia, als veïns, a les nenes i als amics del IQS, els vull donar les gràcies per fer-se sentir a prop malgrat la distància i, cada cop que torno a Barcelona, fer-me sentir com si mai hagués marxat.